ANFN

592.64

# THE EARTH MOVED

Also by Amy Stewart

*From the Ground Up: The Story of a First Garden*

# THE EARTH MOVED

*On the Remarkable Achievements of Earthworms*

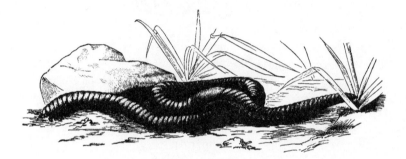

by AMY STEWART

To Dr. Moore —

enjoy!

A. Stewart

ALGONQUIN BOOKS OF CHAPEL HILL 2004

Published by
ALGONQUIN BOOKS OF CHAPEL HILL
Post Office Box 2225
Chapel Hill, North Carolina 27515-2225

a division of
Workman Publishing
708 Broadway
New York, New York 10003

Library of Congress Cataloging-in-Publication Data
Stewart, Amy.
The earth moved : on the remarkable achievements
of earthworms / by Amy Stewart.
p. cm.
Includes bibliographical references (p. ).
ISBN 1-56512-337-9
1. Earthworms.   I. Title.
QL391.A6S733 2004
592'.64—dc22                              2003052379

10   9   8   7   6   5   4   3   2   1
First Edition

To PSB

He thought of all the infinitesimal motions of the world, the obstinate, heartbreaking progress of an earthworm, eating its own route forward.

—CARRIE BROWN, *Rose's Garden*

# TABLE OF CONTENTS

## Author's Note

Readers of this book will quickly discover that I am not a scientist, just an ordinary gardener who is curious about earthworms. My inquiry into the habits and lifestyles of worms brought me into contact with dozens of biologists, botanists, and taxonomists, all of whom went to great efforts to explain their complex research in terms that even I could understand. With their help, I struggled through technical papers and biology textbooks that would have otherwise been utterly beyond my grasp. Any errors, omissions, or failures are my fault alone; when I did get it right, the oligochaetologists deserve all the credit.

I use the words "worm" and "earthworm" interchangeably throughout the book; in every case I refer to terrestrial worms, those creatures that belong to the taxonomic class Oligochaeta. Cabbageworms, cutworms, parsley worms, and tomato worms are not worms at all but moth or butterfly caterpillars. Roundworms, tapeworms, flatworms, and ribbon worms are interesting creatures in their own right but are not the subject of this book.

# PROLOGUE

THERE IS A DIAGRAM of an apple tree pinned to the wall above my desk—an entire apple tree, meaning that the drawing shows its roots as well as its trunk and branches. The tree itself is only five or six feet tall, but the roots extend an astonishing twelve feet into the soil and spread much wider than the outer boundary of the tree's canopy. What's fascinating about the drawing is this: the part of the plant that we think of as the apple tree is, in fact, a fairly insignificant part of the full plant. It's just a squat, knobby protrusion at the top of a graceful, expansive system of roots.

Or is the tree at the top of the drawing at all? In some ways, the tree really seems to be at the bottom of its enormous root system. When I turn the picture upside down, so that the roots are on top and the tree is underneath, a much more graceful creature emerges. The limbs run like rivers in every direction. The shape of the root system is perfect, as airy and symmetrical as any arborist could hope to achieve through years of careful pruning.

When the drawing is turned upside down like this, I am forced to think about the tree's function in a different way. The branches and leaves and fruit are significant, of course: they provide pollen for honeybees, branches for nesting birds, fruit for the gardener, and leaves to carry on the endless respiration of oxygen into the air. But now that I've taken a second look, I see that the roots are the real body of the tree, and I wonder, in a way that perhaps I've never wondered before, what kind of life those roots have underground. How far does the rainwater penetrate? What does the earth look like below the surface? If you asked someone what the ocean is like below the surface, most people could give you a reasonably accurate description. But how little most of us know about life belowground, even in our own backyards.

I realized that I understood very little about the plot of land under my own house. Do I even hold title to this ground twelve feet down? What about twenty, fifty, a hundred feet? The earth's crust extends about fifteen miles down, here on the coast where I live. Beyond that, the mantle is thousands of miles thick. Is this little piece of earth mine, all the way down to its hot red center? Surely at some particular depth I lose my claim to it, and it becomes part of a vast unexplored territory owned by no one.

And who lives down there, under my house? When I think of my property as extending not just across to the neighbor's fence, and back to the alleyway, but down a hundred feet or more, I realize that I paid a paltry sum for a kingdom that just happened to have a house sitting on top of it. Millions—no, billions—of organisms inhabit my little piece of land, and it shocks me to realize how little I know of them.

The first inhabitant of the soil to capture my attention was an earthworm. I am a gardener, after all; I can't miss the fact that gardeners and earthworms work in tandem, tilling the soil, feeding the plants. Still, I've always suspected that there was something more to the story of earthworms. I thought they might have a few surprises in store for me, so I began investigating their habits. I soon realized that they held the key to most of what was happening belowground.

I KEEP THIS PICTURE of the apple tree because it reminds me of something else: a plant's real beauty, its true purpose, might not lie aboveground in the tiny dominion of my garden. There is more to an apple tree than what we can see, much more. To know the land for what it is, to find its heartbeat, to expose its soul, you have to go underground where it lives and breathes.

# Darwin's Worms

It may be doubted whether there are many other animals
which have played so important a part in the history of the
world, as have these lowly organised creatures.

—CHARLES DARWIN, *The Formation
of Vegetable Mould, Through the Action of Worms,
With Observations on Their Habits,* 1881

THE FIRST TIME I held a worm in my hand, I was surprised at how light it was, how harmless. It didn't slither around
or try to get away. Instead it lay curled in a near-perfect circle, as
if it had already accepted its fate.

The worm I held was a red wiggler, Latin name *Eisenia fetida.*
It is in many ways a quintessential worm, small and reddish
pink, with faint stripes between each segment. It is a master
composter, preferring a heap of rotting garbage to just about
anything else. Dig around in pig slop, barnyard manure, or a
mound of damp leaves, and you'll probably find red wigglers,
eating and laying cocoons in the mess. But the worms themselves are not messy; this one slipped out of its pile of rubbish
perfectly clean.

It came out of my worm bin, a small composting operation on my back porch in which I deposit scraps from the kitchen. I don't know how many of them live in the bin—ten thousand, maybe. Sometimes when I dig around in there, the worms are so thick that they look like ground beef set in motion, a mass of churning bodies. It is hard to think of them as individuals, but when it came time to pull one out of the bin and set it on my palm, I did spend a minute looking down at them, trying to choose the right one. A good sturdy specimen was working its way up the side of the bin as if it was ready for adventure.

The reason I was choosing a worm to hold was that it had occurred to me that in all the years I've kept them, I'd never actually touched one. Strange that I would have such an aversion to letting one get next to my skin. How was I to learn anything about the dark and damp place where the plants in my garden put down roots, if I wasn't ready to get intimate with an earthworm?

With one finger, I poked at the worm in my hand. It was completely limp. I could see a purplish vein running along the length of it, just beneath the skin. I curled my palm around the worm, folding it in half and in half again. It didn't react. I began to wonder how a creature this weak could do anything, even move through dirt. Then a few seconds later, it seemed tired of this expedition. It raised one end up—the head, I suppose—and extended one segment at a time into the air. Now, finally, it moved and left a little slime in my palm. I shuddered but didn't drop it. This slime, this worm mucus, was its way of reacting to stress—stress that I had brought on by pulling it out of its bedding and exposing it to light. The worm moved to the edge of my hand, and this time pointed its head down towards the bin, towards home. It was intent on getting back. Just then it looked as if it were capable of doing something after all. It moved with purpose, seeking to escape, trying to return to its familiar habitat. I dropped it into the bin, where it ducked under a layer of damp newspaper and disappeared.

I held worms quite often after that—not just from the bin, although I did get into the habit of pulling four or five out at a time and letting them wiggle through my fingers. I also started handling worms I found in the garden, particularly the enormous nightcrawlers, *Lumbricus terrestris,* that stretched the length of my hand. A nightcrawler, I learned, would press its tail against my wrist as if it was seeking traction, then point its head out beyond the end of my middle finger. On a rainy day, I might handle a half-dozen nightcrawlers like this. It is both fascinating and a little disturbing to pull something out of the ground and stare at it, something that does not belong up here with the rest of us.

WHEN I STAND OVER a patch of earth and wonder about the subterranean activity taking place underfoot, I am not alone. Gardeners are inquisitive by nature; we are explorers; we like to turn over a log or pull up a plant by the roots to see what's there. Most of the gardeners I know are, like me, quite interested in earthworms, in the work they do, churning the earth, making new dirt. We hold soil in our hands, squeeze it and smell it as if we are checking a ripe melon, and we sift through it to see what inhabits it. Ask a gardener about the earthworm population in her garden, and I guarantee she will have something to say on the subject.

It seems strange, then, that most scientists before Charles Darwin didn't consider worms worthy of study. Very little was known about them in the nineteenth century, when Darwin emerged as a sort of champion of worms, devoting his last book to a painstakingly detailed research of their physiology and behavior. *The Formation of Vegetable Mould, Through the Action of Worms, With Observations on Their Habits* was published in 1881. He was an old man when he wrote the book, but the subject had interested him for decades.

How could such an insignificant creature capture the attention of a distinguished scientist like Darwin? He knew from an early age that earthworms were capable of far more than most scientists gave them credit for. He recognized, in a way that no scientist before him had, that they possessed an ability to bring about gradual geological changes over decades, even centuries. This notion—that the smallest changes could result in enormous outcomes—fit perfectly with his work on evolution and the origin of species.

The story of Charles Darwin and his worms begins in 1837, when Darwin was not yet thirty years old. He'd just returned from a trip around the world on a British sailing ship called the *Beagle*. He was offered passage because the captain, Robert FitzRoy, wanted a gentleman on board who could share the captain's table with him. The boat would travel to the coast of South America, where Darwin would have ample opportunity to do the work of a naturalist, collecting specimens and recording his observations. Darwin could not resist the opportunity—he'd only just then been trying to find a way out of the career path his father had laid down for him, that of a parson in a country parish where young Darwin would have plenty of time to chase butterflies and beetles in between his duties to the parishioners. It was not the ideal career for the man who would come to be known as the father of evolution. As one biographer put it, "There was, needless to say, the small matter of his faith." A journey around the world would put off these troublesome questions for a while, and his father agreed to the expedition. But once on board, Darwin realized that it would not be the idyllic adventure he had hoped for: the crew encountered more than its share of dangerous weather, the captain suffered some sort of breakdown midway through the voyage, and Darwin himself was often sick and discouraged. Still, he worked steadily, collecting artifacts and taking notes.

He was away for five years, longer than he could have predicted, and he came back with far more new discoveries than he could have imagined. He arrived in port with over two thousand journal pages, fifteen hundred preserved specimens, and nearly four thousand skins, bones, and dried specimens. It would take him years to organize it all, and even longer to realize the full impact of what he'd collected, for it was here, in this collection of fossils and insects and bird skeletons, that he would begin to see the patterns that would point him towards a theory of evolution. The idyllic vision of a country parsonage was long forgotten. Darwin had now chosen a life for himself as a scientist.

But this was no easy path, and there was no steady employment for a man of his talents. He arrived home from his *Beagle* voyage exhausted, overwhelmed by the work that lay ahead of him, and uncertain of his future beyond that. At first he worked furiously on his collection of notes and field journals, but it was not long before his health was so compromised that friends persuaded him to spend a few weeks in the country. He traveled to Shrewsbury to recuperate at the home of his uncle, Josiah Wedgwood. Upon arriving at Wedgwood's home, he scarcely had time to set down his hat before his uncle had him out in the pastures, where he pointed to cinders and pieces of brick that had been spread on the ground years before and had since become buried some inches beneath. Wedgwood was convinced that the objects had become buried through the actions of earthworms, a feat that would require far greater strength and single-mindedness of purpose than had previously been believed possible of the lowly worm.

In spite of all that he had seen on his voyage around the world, Darwin was impressed with the discovery that his uncle had made in his own backyard. Darwin made a presentation on the subject to the Geological Society of London later that year. At the time, scientists were asking such seemingly simple questions as, Where does dirt come from? Why does dust fall on ships at sea? (Darwin addressed the latter question in a paper that he called, in his typically straightforward way, "An Account of the Fine Dust Which Often Falls on Vessels in the Atlantic Ocean.") After his visit to his uncle's home, he began to believe that earthworms, and earthworms alone, were responsible for the rich uppermost layer of soil, which was, in his day, referred to as vegetable mould.

Although he made some revisions to his first paper on earthworms and saw it published in the Geological Society's journal again a few years later, by this time he was focused on the publication of the account of his voyage on the *Beagle,* and he'd already begun a number of other projects, including the manuscript that would become *On the Origin of Species*. Over the next few decades, he would publish books on the habits of climbing plants, the expression of emotions in humans, the fertilization of orchids by insects, and the variations among domesticated animals. During that time, he would also continue to revise his most well-known works, *The Descent of Man* and *On the Origin of Species*. If earthworms occupied his thoughts during those years, they did not make much of an appearance in his published writings.

Still, when he returned to earthworms in his old age, the book he wrote on the subject would prove surprisingly popular. "As far as I can judge, it will be a curious little book," he wrote prior to publication of *The Formation of Vegetable Mould*. "The subject has been to me a hobby-horse, and I have perhaps treated it in foolish detail." Nevertheless, the book attracted nonscientific readers who enjoyed its clear and vigorous writing and its surprising conclusions.

He described the volume of soil that earthworms swallow and eject as castings, or earthworm manure, reporting that an acre of garden soil could contain over fifty thousand earthworms and yield eighteen tons of castings per year. He studied earthworms' ability to bury objects in soil, from handfuls of chalk scattered on the ground to Roman ruins that had, he believed, come to be buried and preserved for archaeologists by an industrious earthworm population. Most of all, though, he credited them with the transformation of the soil itself. "Their chief work is to sift the finer from the coarser particles, to mingle the whole with vegetable debris, and to saturate it with their intestinal secretions . . . no one who considers the facts . . . will hereafter, as I believe, doubt that worms play an important part in nature."

At the time, people thought his estimates were grossly over-inflated and his claims exaggerated. No scientist before Darwin had taken such an interest in the creatures living underfoot. Earthworms were still largely considered a garden pest that damaged plant roots and spoiled clean green lawns with their castings. At best, they were thought to provide some small ser-vice by perforating the earth and allowing water to penetrate. At least one reviewer of Darwin's early papers insisted that they were too small and weak to carry out the massive movements of soil to which Darwin assigned them. Another critic dryly ob-served, "In the eyes of most men . . . the earthworm is a mere blind, dumb, senseless, and unpleasantly slimy annelid. Mr. Darwin undertakes to rehabilitate his character, and the earth-worm steps forth at once as an intelligent and beneficent per-sonage, a worker of vast geological changes, a planer down of mountainsides . . . a friend of man."

Darwin wasn't deterred by the criticism of his colleagues. "The subject may appear an insignificant one," he admitted, "but we shall see that it possesses some interest." He could hardly re-strain himself before laying out his central thesis: his remarkable conviction that "all the vegetable mould over the whole country has passed many times through, and will again pass many times through, the intestinal canals of worms." It is a stupendous achievement for a blind and deaf creature with no spine, no teeth, and a length of only two or three inches. Scientists of the day could scarcely believe it, and they were quick to express their skepticism.

Darwin had heard these criticisms before in response to the earlier paper he had presented to the Geographical Society, and he did not waste the opportunity to both refute his critics and remind them whom they were up against. After all, he'd fought most of his life to win acceptance for his theory of evolution, and he saw parallels between his work on evolution and his work with worms.

A scientist looking back over Darwin's work wrote that "the key to his genius was the ability to stretch his imagination to encompass geological time—thousands of years, thousands of centuries." He understood that tiny, incremental changes in the environment could bring about the evolution of a species. It was this same approach that led him to understand that soil could, over time, be transformed through the efforts of earthworms.

"Here we have an instance," Darwin wrote of his detractors, "of that inability to sum up the effects of a continually recurrent cause, which has often retarded the progress of science, as formerly in the case of geology, and more recently in that of the principle of evolution." He dispatched a French scientist who disagreed with his conclusions about the abilities of earthworms, making the calm statement that the Frenchman "must have thus argued from inner consciousness and not from observation," for Darwin's own observations bore out the truth. The power of earthworms, then, came not from their individual, but from their collective strength. It is a surprisingly egalitarian conclusion to reach about earthworms, one that could only come from a man who had great vision and also great affection for the creatures themselves.

Today, among earthworm scientists, Darwin is a kind of touchstone, a muse. He looked belowground with real interest and treated the dark earth as the mysterious unexplored territory that it is. He lived at an exciting time for scientists: in every corner of the world, exotic plants and birds and fossils awaited discovery. But he chose to look underground, to seek out the earthworm. Now we know that Darwin had only glimpsed the potential power of worms: his conclusion that over fifty thousand worms could inhabit an acre of soil was in fact quite low. Scientists have shown that figure to be one million. Earthworms in the Nile valley can deposit up to a thousand tons of castings per acre, helping to explain the astonishing fertility of Egypt's agricultural land. As Darwin had only just begun to suspect, earthworms pass the top few inches of soil through their guts every year. This makes them beings to be reckoned with, a force for change in more ways than even he could have guessed.

Over the last one hundred years, earthworm scientists (called oligochaetologists, named after the taxonomic class in which earthworms fall, Oligochaeta), have come to quantify what farmers have always known: that worms, through their actions, substantially change the earth. They alter its composition, increase its capacity to absorb and hold water, and bring about an increase in nutrients and microorganisms. In short, they prepare the soil for farming. They work alongside humans, extracting a life from the land. They move the earth, a remarkable accomplishment for a creature that weighs only a fraction of an ounce.

• • •

AN EARTHWORM TRAVELS through the soil, pushing some particles aside and ingesting others. Although its food choices may look alike to the casual observer, the worm is actually sorting through the soil in search of tiny bits of decaying organic matter, which it will swallow along with some clay or sand particles. It builds a permanent burrow as it goes. At night it rises to the surface of its burrow, ejecting a small mound of castings around the entrance. It searches for food, tugging leaves, pine needles, and other detritus into its burrow. This simple routine is enough to ingratiate it to the farmer or gardener. On its nightly forage for food it acts like a small, very efficient plough.

The body of an earthworm is perfectly designed for life underground. Sight is unnecessary in the subterranean world; a sensitivity to light is all a worm needs to avoid straying out of its habitat. Lungs are not much use in the tight confines of a burrow; instead, the earthworm breathes through its skin, exchanging oxygen for carbon dioxide, relying on damp conditions to help absorb the oxygen in the same way that the damp interior of a mammal's lungs facilitates the passage of air into the body. The earthworm's shape allows it to be an extraordinary vessel for soil—the perfect container for holding, transporting, and transforming earth.

"The plough is one of the most ancient and most valuable of man's inventions; but long before he existed the land was in fact regularly ploughed, and still continues to be thus ploughed by earth-worms," wrote Darwin. Although he studied many aspects of earthworm biology and behavior, the august scientist was especially intrigued by its ability to sift the earth. He watched them emerge from their burrows at night and draw in twigs and leaves or even drag small stones over a gravel walk until they formed a pile at the mouth of the burrow. He crept outside and unplugged enough of these burrows to know that the worms rested just inside, their heads readily visible just below the surface. Were they hiding from predators? Trying to keep rainwater out? Perhaps they were just protecting themselves from the cold night air. Whatever the reason, this nightly gathering of materials and systematic drawing in of leaves and plugging of burrows was certain proof of their unlikely physical strength and engineering abilities.

If a person were to pull leaves or twigs into a hole, Darwin reasoned, they would grab the object by its narrowest end and pull it in. If the object was long and skinny like the hole itself—say, a twig or stem—they would probably pull the thickest, heaviest end in first. Surely instinct alone could not account for the manner in which a worm selected material for its burrow. Intelligence, he declared, had to be the guiding factor. When the worms reached for fallen leaves and twigs around their burrows, they were selecting the best material available. They evaluated, they experimented, they made decisions.

Let me say that again: they made *decisions*—actual decisions, made after trying several alternatives and choosing the one that seemed best for the situation. This is perhaps the most surprising revelation in Darwin's book. Although earthworms had undoubtedly been making such decisions for centuries, they had a new and unlikely advocate in Charles Darwin. He had the time, the resources, and the scientific methodology to prove that what earthworms did was more than mere chance.

I THOUGHT OF DARWIN and his worms when I was out in the garden, digging a new vegetable bed for the three dozen asparagus crowns that arrived by mail. A layer of fog had descended over Eureka, covering the hills around Humboldt Bay I can usually see from here. The earth was damp but not muddy, just right for planting.

I pushed a pitchfork into the soil and leaned back on the handle just enough to raise the tines of the fork and disturb the ground. My days of double-digging— of scooping out the top layer of earth and the one beneath it, filling in the trench with compost, and placing a mixture of soil and compost on top— are over. The soil is an intact system, a community of microorganisms that lives and breathes, and it will function best if I don't disturb it too much.

Once the ground was loosened I spread a layer of compost on top. The microbes— the bacteria, the protozoa, the fungi— could work their way into the earth gradually, and the earthworms would rise to the surface and take the compost down with them. Down the center of the bed, I pulled apart the soil with a hand spade and created a narrow trench to bury the crowns. A layer of compost went in the bottom, and then I pulled the crowns out of the box, and spread the roots so they straddled the compost. I knocked enough dirt back into the trench to cover the crowns, but a shallow depression remained. I planned to fill it in slowly over the next few months as the first asparagus shoots appeared. The extra soil around the newly formed shoots would make them pale and tender, at the same time providing enough nutrients to encourage them to grow tall and robust.

There were easily a few dozen earthworms inhabiting the newly dug asparagus bed. Each worm holds less than a teaspoon of earth in its body as it moves through the soil. In a day, they will eat about a third of their body weight in soil, maybe more. This doesn't sound like much, but even Darwin's conservative estimates showed that over the course of a year, a healthy earthworm population can move almost twenty tons of soil per acre.

I leaned against my shovel, calculating that I'd spread about thirty pounds of compost over my asparagus bed. Over the next year I could expect earthworms to add another thirty pounds of castings around the roots of the plants. If conditions are right, they'll supply another thirty pounds—maybe more—the following year, and the year after that. These asparagus crowns will produce for over twenty years. In that time, if the earthworms flourish, they'll contribute about six hundred pounds of nutrient-rich castings to this small space, taking care of my vegetable bed far more efficiently than I ever could.

DARWIN IS RESPONSIBLE for putting these kinds of thoughts in my head. My gardening chores take significantly longer now that I slow down to count worms, now that I sit in the garden path, chin in hand, calculating the volume of castings. I have slowed down, it seems, to Darwinian time. He had that luxury in his later days; he could spend hours out in the fields around his house, watching earthworms and collecting their castings after they had disappeared into their burrows, making guesses about how they spent their time after they vanished from sight.

He also had the good fortune to know scientists around the

world, and those colleagues sent him specimens and castings in the mail. He weighed and cataloged them, made a note about the area where they were collected, and organized the results into tables. Thanks to his meticulous approach, his work today remains some of the best data on earthworm activity. He wrote this in his autobiography: "I think that I am superior to the common run of men in noticing things which easily escape attention, and in observing them carefully."

There is no doubt that he took some pleasure in his work. He had a genuine fondness for the worms and seemed to enjoy the painstaking effort that his research required. I can only imagine that his experiments on their habits were a daily delight in his old age. One biographer wrote that Darwin "became in the end what he had always been in his heart, almost a part of nature himself, a man with time to lean on a spade and think, a gardener." I like to imagine him as a dabbler, a homebody, a man who explores his most intimate surroundings with both deliberation and wonder. In the waning years of his life, he was sometimes weak and infirm, but that only turned the attention of his scientific mind away from the wider world and towards his home, his garden, and the earth.

The approaches he used to evaluate earthworms were, by this time, classic Darwinian methods. Throughout his career, he took an ingenious, almost playful approach to experimentation. Like most naturalists, he was a tinkerer, interested as much in nature's minutiae as in its grandeur. He liked the inner workings, the tiny springs and gears of the natural world. Perhaps he felt that nature's true power rested there, in the movement of pebbles and seeds, and in the commerce of ants and worms.

Think of him in his laboratory, with his notebooks and specimens. One day he becomes interested in the mechanism that allows climbing vines to climb, and he ties small weights to the tendrils of plants to see how they respond. They hang on the vines like miniature Christmas tree ornaments, forcing the plant to reveal its tricks. He marvels at plants whose leaves roll tightly shut after dark. How could a plant act so deliberately, with such intent? He forces their leaves open so that they cannot close at night, hoping to lay bare their secrets.

And now, when the old man turns his attention to worms, picture him stealing outside on wet mornings to pull leaves out of burrows and observe how they had probably been tugged inside. He gathers a handful of pine needles and scatters them around burrows to see how worms will handle them. Eventually his curiosity about their mental capacity leads him to cut out irregularly shaped paper triangles and set them among active burrows, then chart the number of times the triangles are drawn in by the apex, the middle, or the base.

Darwin was meticulous with his research. Since this was to be his last book, he seemed determined to get it right, to document every element of earthworm life. He pulled not a few leaves out of burrows; he pulled 227 out and reported that 181 of them, or eighty percent, had been drawn in by their tips. The others had been drawn in by their bases or seized in the middle, causing the leaf to crumple once inside the burrow. The image of the elderly scientist pulling 227 leaves out of burrows and cataloging them to prove the intelligence of earthworms in his backyard is amusing, even surprising, but he didn't stop there; he went on to reconstruct pine needles by breaking them apart and rejoining them at the base using glue or thread. He aimed to prove that worms knew to drag them into their burrows by the base where the needles were joined, rather than by one end, which would surely result in the needle getting stuck midway. He wanted to demonstrate that they were not acting out of instinct, because of a pine needle's particular taste or feel. He created 271 of these artificial sets of pine needles and observed that eighty-five percent of them were drawn in by their bases, noting that worms were slightly more likely to draw pine needles in by the base if they were attached with thread as opposed to being attached with glue, which might have smelled or tasted unpleasant to the worms. He wondered if the worms naturally avoided the sharp points of pine needle ends and chose the base because it was rounder. To test this, he carefully trimmed off the sharp ends and found that worms drew them in by their base regardless.

As for the paper triangle experiment, he did not simply cut a few triangles and leave them lying around. He cut 303 triangles of various sizes, coated them with fat to keep them from going limp in the night dew, and established some baseline data by drawing triangles into small tubes using tweezers to determine the most efficient method that he would employ if he, rather than the worms, were given this task. (He chose the apex, as opposed to the middle or the base.) Even working with this unfamiliar material, they drew the paper triangles in by their apex sixty-two percent of the time. He went on to observe that the triangles pulled by their apexes had been drawn in cleanly, with very little evidence of fumbling around or trial-and-error first. "We may therefore infer—" he writes, "improbable as is the inference—that worms are able by some means to judge which is the best end by which to draw triangles of paper into their burrows."

ONE OF DARWIN'S most extraordinary qualities was his ability to recognize when a scientific question could not be answered due to the limitations of the science of his day. He knew, for instance, that during his lifetme, no significant progress would be made on the question of how life first began. Near the end of his life he wrote to a colleague, "You expressed quite correctly my views where you said that I had intentionally left the question of the Origin of Life uncanvassed as being altogether *ultra vires* [beyond the powers] in the present state of knowledge." The same could be said of Darwin's insight into the role of earthworms in the soil. The technology that would allow scientists to understand the complex relationships between soil microbes, plants, and earthworms would not be advanced for several more decades.

When *The Formation of Vegetable Mould* was published in 1881, it was a novel idea that an earthworm could possess enough intelligence to judge how to best pull objects into its burrow. No scientist had paid so much attention to this seemingly trivial matter, nor devoted so many pages of published work to it. But even Darwin could not grasp the importance of the earthworm's impact on the soil ecosystem. Scientists in his day knew that bacteria and other microorganisms lived in the soil, but the ideas were quite new. Louis Pasteur initiated the science of microbiology during the last few decades of Darwin's life. A strain of bacteria was first identified as the cause of a plant disease in 1878.

Still, the relationship between the microscopic world of soil and the macroscopic ecology—the earthworms and other visible creatures that inhabit the earth—was largely a mystery. Over the next several decades, the study of earthworm behavior was eclipsed by the study of its role in the soil. To understand what's happening underground, we have to know more about this creature that lives below our feet, selectively drawing organic matter down from the surface, creating pockets of air everywhere it goes, sifting and digesting particles of earth.

# Unsung Hero

I spent whole afternoons in the dirt, making my patch of ground flawless. I even cleared the worms away, before I found out that all the tunnels they make give air, and probably other molecules I don't know about yet, to the plants.

—JANE HAMILTON, *The Book of Ruth*, 1988

I HAVE KEPT WORMS on the back porch in a worm composting bin I bought on impulse seven years ago. At the time, I didn't know much about earthworms, having only the vaguest idea that there was more than one species, and harboring nothing more than a dim notion about the value of their castings to my garden.

The bin itself—a recycled rubber contraption called a Can-O-Worms—is both simple and marvelously well designed at the same time, not unlike the earthworm itself. Three round trays stack on top of one another. Each one has a hole on the bottom to allow the worms to move up through the trays. They all nestle into a base, which sits atop three plastic legs and contains a spigot to drain liquid. The worms start out in the bottom tray, where they consume vegetable trimmings, coffee grounds, and newspaper. Once they've eaten their way through the bottom tray and filled it with about three inches of castings, the second tray is added and they move up into it, seeking food. The process is repeated for the third tray. The castings in the bottom tray can be emptied out into the garden, and then returned to the bin, serving as the new top tray. The process repeats itself, the worms moving through the rotation of three trays.

When I bought the worm composter, I was so excited that I could not wait even a few days to order from a worm farm, so I selected my worms from among the Styrofoam containers in the cooler at the bait stand. The red wigglers I purchased turned out to be sturdy, hardworking creatures that, after being spared the fate of the bait hook, were more than willing to spend their days eating my garbage. I probably bought twenty cartons, which works out to roughly one thousand worms. Now, seven years later, they have multiplied again and again. I don't know how many I have—five thousand? Ten thousand? I added a second bin a couple years ago; perhaps between the two I have almost twenty thousand worms living just a few feet from the kitchen table where I drink my morning coffee. By the standards of most worm farmers, this a small operation, just large enough to keep up with the food scraps that one household produces.

When the worms came out of their bait stand cartons, they were less than an inch long and as thin and limp as a strand of spaghetti. They only had to survive long enough to make it onto the end of a fishing hook; no one expected these worms to go to work making compost.

But once I got them settled into the bin, it did not take long for them to fatten up and start reproducing. The next generation was substantially more healthy than the last: they were easily as long as my finger, measured twice as big around as the first batch, and moved vigorously through the bin, churning through potato peelings and old tea bags as if this was the work they'd been born to do—and, in fact, it was.

THESE RED WIGGLERS I'd purchased don't live in the soil. They thrive in manure piles and layers of rotting leaves, where they feed off the bacteria that does the real work of breaking down organic matter. On the other hand, if I had dug up worms from my garden and put them in my new bin, they would have found the environment completely inhospitable. As a keeper of worms, one of the first things I had to learn was that not all worms behave the same way in the soil. Scientists find it useful to classify worms by function, and once I'd learned the system, I could pick them out in my own garden. Small yellowish-brown worms—endogeic worms—turn up only in the roots of plants; I never saw them in an empty vegetable bed. Large burrowing nightcrawlers, called anecic worms, live deep in the soil. Unless the weather had been damp for several days, they were hard to find near the surface. Epigeic worms, like the red wigglers that occupied my worm bin, would make a home in rotten muck on the surface of the ground but never deep in the soil.

My red wigglers didn't rate a mention by Darwin but are among the most studied worms today, showing enormous promise as composters of food and industrial waste. Epigeic worms have a relatively short life span, living just a few years, and are known for their ability to reproduce quickly to match the available food source. (The exact reproductive rates are a source of some controversy. Many budding entrepreneurs have been lured into a kind of earthworm pyramid scheme that promises that a healthy worm population will double its size every sixty days, a growth rate that conjures up images of wildly escalating profits unmatched by any other livestock enterprise.) The fact is that reproduction rates are influenced by climate, environment, and food source. Epigeic worms have been known to lay anywhere from a dozen to a few hundred cocoons in a year. Many of them may never hatch if conditions are not right; others may hatch two or three young earthworms apiece.

Two of the best-known epigeic worms can probably be found at a bait stand in a carton labeled "red wiggler" *(Eisenia fetida)* or "redworm" *(Lumbricus rubellus),* or in a grade-school science classroom, where they often provide the basis for ecology and recycling projects. Both can live in a bin and both are favorites among worm composting enthusiasts. Some people do try to raise them to add to their garden soil, but the fact that they only live in rich, decomposing organic matter, and not underground, makes them best suited for worm bins, outdoor compost piles, or very heavily mulched soil. Most worms in this category are under three inches long and dark red or almost brown, and many have pale stripes between their segments.

The castings that epigeic worms produce are a rich soil amendment. These worms are uniquely well designed for their role in the detritusphere, or leaf litter layer of the soil, where they deposit nutrients that help plants germinate and grow. Here is just one example of an interaction between an epigeic worm and the plant life it supports: the epigeic worm is endowed with an active calciferous gland that helps it process calcium in its diet and excrete the excess in its castings. Calcium is critical to plant growth because it allows plants to take up nitrogen, which promotes leaf growth, and it assists with protein synthesis and other vital plant functions. Anyone who has ever picked a ripe tomato from the garden only to find a mushy brown spot on the bottom—an affliction known as blossom end rot—has discovered what a lack of calcium can mean. Fortunately, blossom end rot is specific to each individual fruit: a plant that has produced one or two spoiled tomatoes may go on to produce healthy ones later in the season. This means that a gardener like me will run to the nursery at the first sign of blossom end rot, hoping to find a calcium supplement that will ensure blemish-free tomatoes later in the summer. Dolomite lime, bone meal, and gypsum are all sold as calcium supplements for home gardeners and farmers alike. Although they've been proven to work against problems like blossom end rot, any off-the-shelf supplement has its own set of drawbacks. Many nutrients leach away during heavy rains, decompose too quickly to be of use to plants, or exist in a form that they can't readily absorb. That's where earthworms, particularly epigeic worms, play such a critical role. They address calcium deficiencies on several fronts, producing calcium in their calciferous glands during digestion, adding it to the soil through their castings, or even transforming it, as it moves through the intestine, into a form that is easier for plants to absorb.

. . .

DARWIN FOCUSED ON an anecic worm called *Lumbricus terrestris*—the nightcrawler—in his book. This is the worm that builds permanent vertical burrows in the soil, leaving little mounds of castings alongside the opening. It can burrow as deep as eight feet, rising to the surface at night in search of food. Unlike epigeic worms, anecic worms like nightcrawlers do ingest some soil along with dead leaves and other decaying matter. They are large, powerful worms well suited to the work of tilling the earth. A typical nightcrawler has a fairly long life cycle —up to six years—and reproduces more slowly than an epigeic worm might. Because it burrows deep into the earth, it can wait out drought cycles. It is considered a somewhat sluggish creature in comparison to its red wiggler cousin, but it can move quickly when it is trying to withdraw into its burrow to escape a predator.

Nightcrawlers are typical of most anecic worms in terms of size and coloring. They are dark red along their dorsal side, or backside, and are nearly translucent on their ventral, or underside. Their rear ends are flattened in what scientists call a spoon-shaped posterior. The clitellum—a thickened band of skin about a third of the way down their body—is quite pronounced, and with some magnification, one can even see the tiny bristle-like hairs called setae that worms use to anchor themselves in their burrow or to hold onto one another when they mate. Nightcrawlers are, in some ways, the archetypal worm, instantly recognizable to anyone who has ever picked them out of the lawn for bait or watched a robin pull one from its burrow, stretching it out of the ground while its spoon-shaped posterior clung to the burrow with every seta it possessed.

I first encountered *Lumbricus terrestris* in my own backyard. It turned up on the end of my shovel, six inches long, perfectly pink and clean in spite of the fact that it had lived its entire life in the dirt. I knew that the nightcrawler was a common enough worm in most parts of the country, but I never saw one in my first garden back in Santa Cruz. It had its share of earthworms, but none as large as the nightcrawler.

Anecic worms are valued not just for the castings they produce, but also for the ways in which they move soil around. Darwin claimed that earthworm burrows benefit the earth by letting in air and rainwater. "They allow the air to penetrate deeply into the ground," he wrote. "They also greatly facilitate the downward passage of roots of moderate size; and these will be nourished by the humus with which the burrows are lined." He was, as usual, quite prescient in his description. Oligochaetologists working decades after Darwin's death have discovered that the walls of those worm burrows, called the drilosphere, are rich in bacterial and fungal growth, thanks to the mixture of the particular mucus that the worms excrete and the casts left behind as they move through the soil.

Try pressing your finger into compacted clay earth and see how far you get. Some of the dirt in my own garden is so hard that I can barely make a dent with one finger, and even a shovel has trouble penetrating except on very damp mornings. Now imagine a worm, that limp and spineless creature, working through the same soil. First it anchors its setae into the soil to brace itself, then it stiffens the muscles that run in segments around its body. It increases the pressure inside its coelomic cavity, a vessel that holds the mucus it excretes during locomotion, reproduction, or a time of stress. The increased pressure in this cavity propels the head of the worm forward. It eats a little soil as it moves. The tail contracts, pulling it in the direction its head was moving, and the process begins again. It can take a worm weeks to build a system of burrows in a laboratory study, where the soil may not be as compacted as those unyielding, undisturbed areas in my backyard. Although it is difficult to clock a worm's speed in the wild, we do know that nightcrawlers only migrate a few yards per year. *Lumbricus terrestris* does not race through the soil; it meanders, sticking close to food sources and always seeking damp, cool ground.

OF THE THREE MAJOR categories of earthworms, endogeic earthworms are probably the least familiar to most people, for good reason: they rarely come to the surface. Many endogeic species inhabit the rhizosphere, the area immediately around plant roots, where they feed on soil that has been enriched by decaying roots, bacteria, and fungi. *Aporrectodea caliginosa*, often called a grey worm or a southern worm, is one of the most widespread endogeic species. The name actually refers to a cluster of closely related species that inhabit farmland and pastures. As its common name indicates, it is grey or slightly pink and about two or three inches long. I often turn up a few *Aporrectodea caliginosa* when I'm transplanting shrubs or pulling weeds. Worms in this category are mainly geophagous, meaning that they feed almost entirely on soil, although they do seek out earth that is relatively higher in organic matter. Because they stay below the surface, they are less likely to be harmed when agricultural fields are tilled or when the ground is disturbed during planting. That makes them a popular choice for farmers who want to inoculate their soil with earthworms. But because it is difficult to breed them in captivity, farmers are often advised to find an area of undisturbed pastureland and dig out a square of sod that seems well populated by grey worms. The sod can then be transplanted directly into agricultural fields, where the worms will, over a period of years, colonize the land.

These small, greyish worms aren't the only endogeic worms out there. Much larger ones exist well below the surface of the earth; they are large pale creatures that almost never see the light of day. They can build tunnels ten feet belowground, where they may encounter tree roots but few other plant roots. These deep-burrowing worms are among the only living creatures that can survive at such a depth. Beyond that, only microscopic organisms like bacteria are found, and they disappear almost two miles below the earth's surface, where temperatures reach 160 degrees and life in any form is difficult to support.

Deep-burrowing worms, then, inhabit a world that would seem stark and barren to us. For their size, they are surprisingly delicate. One such worm, *Megascolides australis,* can grow to several feet in length but its skin is so fragile that it could burst if it is handled too much. Its tunnels are so large and well lined with coelomic fluid that some Australian farmers can hear a gurgling sound coming from deep within the earth when the worm is on the move. Another giant worm in Oregon, *Driloleirus macelfreshi,* measures two or three feet long and is known for its coelomic fluid, which smells distinctly of lilies. Both of these species survive only in undisturbed habitats; both are nearing extinction due to the encroachment of cities and roads. They are sensitive to vibrations on the surface of the soil and can detect a bulldozer at work. They move quickly through their burrows to escape notice, making it nearly impossible for scientists to collect specimens or study them in the wild. Still, they can live for many years, possibly decades.

Recent efforts have been made to locate the giant Oregon worm, but none were found. Unless one turns up by accident, scientists have no way of knowing if they have managed to adapt to the industry, noise, and pollution created by the encroachment of humans, or if they are, by now, entirely extinct.

IT HAS TAKEN over a hundred years for scientists to put together this portrait of earthworm's life. Now what emerges about life underground is an image like a map of a city in which the inhabitants build roads as they go, choose neighborhoods based on the abundance of food and availability of damp, dark quarters, and carry microorganisms in their guts like passengers on a bus.

Picture, once more, that drawing of the apple tree with its roots extending twelve feet into the soil. How many earthworms are at work in the roots of that tree? Dozens? Hundreds? An apple tree can live for decades. What difference could a population of earthworms mean to its health and longevity? I am starting to believe that it could make all the difference in the world. In an age of ecological uncertainty, when natural habitats are disappearing and creatures like the giant Oregon earthworm are becoming extinct before they have even been properly identified, when farmland is being paved over in favor of neighborhoods and shopping centers, and farmers are turning to genetic engineering and biological pesticides to solve their agricultural problems, in this complex age, the earthworm may emerge as a kind of unsung hero, one whose potential we are only just beginning to understand.

# The Earth Moved

One could state as a hypothesis that earthworms cannot
survive or do not have natural means of transport across bodies of
salt water. This is testable, although it would be difficult to release
test worms or cocoons by all possible means of conveyance
(rafting vegetation, logs, on debris in violent cyclonic storms
and so on), and even harder to track their fates.

—Samuel James, *Earthworm Ecology*, 1998

Darwin faced a particularly vexing problem
during the development of his theory of evolution. He argued
that a species can change to adapt to its environment, which is
why variations among species could be found in different parts
of the world. There were notable differences among the finches
he encountered in the Galapagos Islands during his *Beagle* voyage in 1835, for instance. The popular thinking of the day, however, was that each species was God's creation, pure and simple.
The fact that near-identical animals could be found on far-
flung continents around the world only supported this notion.
After all, if a species was supposed to change in response to its
environment, how could the same flower, insect, or fossil be

found in Australia and South America? Surely only divine creation could explain it.

Spiritual matters were often at the heart of Darwin's internal conflicts about his work. He took great risks by challenging the notion of creationism as it was understood at the time. Scientists like Galileo and Newton, exploring the laws of motion before Darwin's birth, had made a case that there were, in fact, physical laws that governed the movement of the planets and the stars, as well as the velocity of hail falling from the sky or the speed of a pendulum's swing. These notions were made more palatable by the belief that God, if not directly responsible for moving heavenly bodies through space, at least created the physical laws governing their movement. God, people could believe, invented the laws of physics, then put them to work. He placed the planets in their orbits, and sent them spinning through space with a puff of heavenly breath.

Darwin struggled with a similar notion. Perhaps God did not intervene directly in the formation of each new creature, or each variation within a species. Perhaps God created the laws of evolution and put them to work on Earth. These questions were no small matter to Darwin. A great deal was at stake. His wife, Emma, was a deeply devout woman and he could not bear the thought of causing her—and her family, the Wedgwoods, with whom the Darwins had many family ties—the shame and public ridicule that his ideas could bring all of them. In fact, for a while he considered hiding his essay on evolution that would eventually become *On the Origin of Species* and even planned to leave instructions for it to be published after his death. It was not until other scientists began to reach similar conclusions that he felt his work on evolution could be published without bringing ruin on his family.

Still, for years Darwin tried to come up with an explanation—apart from that of divine creation—for the similarities between species on separate continents. Eventually he settled on the notion that identical species could travel to another continent on their own by wind or water. He set out to prove that seeds could be transported across the ocean by soaking them in salt water and germinating them afterwards. When his experiments showed that they would sink in salt water and might not float across the Atlantic as he had originally thought, he moved on to birds. He fed them seeds, then killed and dissected them to retrieve the seeds from their guts and germinate them. He even asked his colleagues for specimens of bird feet caked in mud to demonstrate that seeds could travel across the ocean in this way, and half-rotted bird feet did indeed arrive by post for his inspection.

Although some mapmakers and geologists had noticed as early as the seventeenth century that the continents of the world had parallel coastlines that might have been connected at one time, Wegener's theory of continental drift was not published until 1912, some thirty years after Darwin's death. Finally it became obvious that the striking similarities in flora and fauna Darwin saw on his voyage were not, as he feared, damning proof against his theory of evolution. Instead, the presence of like species on different continents could be explained by the very notion that had floated around for centuries before: these continents once fit closely together, like pieces of a puzzle. Even earthworm species follow this pattern, telling the story of the movement of the continents in a way that birds and snakes and crickets cannot. Worms, because they live in the earth, are uniquely qualified to document the movements of land masses.

TRACING EARTHWORMS BACK through the earth's history is not as easy as it might seem. There are enough similarities between worm species around the world to suggest that they were well established before the continents started drifting apart. But a soft-bodied animal like a worm or a jellyfish does not leave much of an imprint in the fossil record. Worms do make their mark in the earth; occasionally a fossil will turn up with holes running through it that look distinctly like burrows. And fossil records indicate that earthworm ancestors — soft-bodied, water-dwelling creatures that could have been similar to leeches or marine worms — lived in the Cambrian period, around five hundred million years ago. But when did worms first establish themselves on land? One thing is certain: before there were earthworms, there was soil.

The sea level oscillated enough during the Carboniferous period, about 350 million years ago, to form limestone, shale, and sandstone in what would later become North America. Coal-forming swamps contributed to the vast deposits of coal in England. As the oceans advanced and receded, one layer of sediment after another was left behind. In this environment, ferns and early trees flourished. Here, geologists presume, in the roots of these primitive plants, earthworms would have found their niche. By this time, animals were laying hard-shelled eggs, early reptiles were living in the forests, and land snails appeared for the first time, as did millipedes, scorpions, and spiders. Life underground was perhaps not so different from what it is today. Springtails and mites lived in the soil, alongside earthworms, as they do now. And even in this modern age, the habitats of earthworms harken back to their watery beginnings: you won't find a worm in the desert, and you won't find one under a glacier. They continue to seek damp, cool soil, as they did in their earliest days.

It is difficult to say with absolute certainty how widespread earthworms might have been when, 248 million years ago, the Permian Mass Extinction wiped out many land-dwelling species and all but about five to ten percent of marine life. The causes of extinctions are in general a source of controversy, but glaciers, volcanic eruptions, and the shifting and grinding of land masses are all considered viable theories. This much is clear: the earthworm persevered. No climate change, no geologic upheaval, has managed to threaten its existence.

What came after—the Triassic and Jurassic periods—is perhaps one of the most familiar times in the Earth's ancient history. Dinosaurs appeared, the first birds took flight, and flowering plants evolved. (Another mass extinction about sixty-five million years ago wiped out the dinosaurs and about three-quarters of all species living on the planet at that time, but once again, earthworms survived.) These mass extinctions were not, however, the most dramatic events in the earthworm's history. Something else was happening during the age of the dinosaurs. About two hundred million years ago, a fissure started to form between what is now Africa, North America, and South America, and the Pangea supercontinent began to break apart. Earthworms living on those continents today are so much alike that their very similarity confirms that they once lived together on the Pangea land mass.

THE MOVEMENT OF PANGEA, then, held the answer to Darwin's questions about the similarities of species around the globe. The fossils, plants, and birds that he found on one continent could be closely related to those on another continent. While he had envisioned birds flying across the ocean with seeds embedded in their feet to explain the similarity between species, we now know that it was not the seeds or the birds that traveled so far. The continents themselves have moved, carrying their inhabitants with them.

It is this very issue that has attracted the attention of Sam James, one of the world's leading earthworm taxonomists. He realized that if he could map the distribution of earthworms around the globe, he could establish their place in the study of continental drift. The connection between earthworm taxonomy and geology has not been, in his opinion, adequately explored.

"Take the worms you've got in your California redwood forests, for example," he told me on the phone one day. "They're not very well known, but what we do know about them tells us that their closest relatives are in Australia and New Zealand. How did they get there? Did they wander around the Pacific Rim? Some people say they did just that. On the other hand, there are chunks of eastern Australia embedded in British Columbia. California, Oregon, Idaho, Nevada—they're made up of sutured island areas and ocean floor crust, all pasted together. And it's still moving, at least in geologic time. Sit there in Eureka long enough, you'll see southern California drift by eventually. And what about Africa? We were connected to Africa at one point, but we don't have any worm species in common. Why? Maybe because that area was desert at the time.

"Or look at the worms in the Caribbean. They're closely related to worms in Fiji. How did that happen? That's what I want to find out. You see all these papers published about birds of the Caribbean islands, but you know what? Birds can fly away. Worms can't get across bodies of saltwater by themselves. If you find a native worm in Mexico that's related to a worm in Australia, the only possible explanation is the movement of land masses. That's one area where earthworm science is headed next. Land area patterns. The relationship of worms to the movement of the continents."

Sam realizes that those relationships can be traced only if we can discover how many species of earthworms live on the planet and where they are. But in some ways, the earthworms inhabiting the globe today are almost as elusive as those that lived hundreds of millions of years ago. Most oligochaetologists start out in another field—forest ecology, biology, botany—and discover along the way that earthworms, which play such a vital role in the ecosystem, are not well classified or understood. In Sam's case, a semester-long project to investigate their role in a particular prairie turned into a career.

"I started out studying grasslands," he told me. "Big animals. Cows, buffalo. I was doing research on whether a cow's spit makes grass grow. Things like that." A few years after he began this work, a herd of bison was going to be reintroduced to the same kind of prairie their ancestors had once inhabited. "I got myself invited along," he said. "As soon as they released the animals, I ran out to the paddock and collected all the dung I could carry. I was looking at rapid nutrient cycles. You know, the bison eat the grass, turn it into manure, then the earthworms eat the manure, and they help the grass to grow, which the bison then eat. Then I found out that nobody knew what kind of earthworms were living out there. That's how it all started."

Sam has been on several worm-hunting trips to the Philippines and was eager to tell me about the dozens of undiscovered specimens he found and brought back with him. "One is deep indigo blue," he said. "Eighteen inches long and about a thumb's width in diameter. It's got big white spots with yellow centers, like fried eggs, all over its back, if you can imagine that. And get this—it crawls on the forest floor, doesn't burrow in the ground, and its infants live in trees until they're mature. Amazing."

He showed me this worm one time. He pulled it out of a jar so I could get a better look. The preservation process had faded it a little—he had to hold up a photograph of the living specimen so I could appreciate the iridescent blue-green sheen of its skin—but it was still remarkable. It was like nothing I'd ever seen or imagined, like something out of a science fiction movie. I didn't touch it, but I imagined that it had the texture of a pickle, something that was once pliable and alive but had since stiffened in its brine. It looked more like a small snake than a large worm, except that it had segments running in rings down its body instead of scales. Sam explained that the spots on each worm were different, which is unusual for a creature that is generally symmetrical. "If you were to observe these in the wild somehow," he said, "you could get to know them by their different spot patterns, the way people learn to tell individual whales or birds apart.

"Down in the Philippines, they have all these posters, you know, like 'Birds of the Philippines,' 'Butterflies of the Philippines,' that sort of thing. Well, we're making a 'Worms of the Philippines' poster. They've got a lot of remarkable worms over there. People don't realize that."

Sam James is passionate about discovering and naming unusual species. "I've got so many new worms here, I could spend years identifying and classifying them. I'm thinking about putting them all on a website. People can get a worm named after them, or you could get one named for your husband for an anniversary present. You know, like they do with stars. What do you think?"

What do I think? It's brilliant, I tell him. Everyone will want one.

WITH SO MANY undiscovered species, it's a wonder that more scientists haven't picked up a shovel and gone digging for them. Who could help but be entranced by the notion of an indigo worm with yellow and white splotches on its back? Yet there are surprisingly few scientists working on earthworm taxonomy. Canadian scientist John Reynolds is one of the few who has spent a lifetime working in this field. The work is incomplete and, as he told me, terribly underfunded.

"Most worm specialists have to work a day job," he said. Right now he's in his ninth career, at a Canadian trucking firm. "I've also been a silversmith, a research scientist, a professor, a—oh, let's see—a lawyer, a police officer, a college dean, a consultant, then I was a truck driver, and now I'm in what you'd call transportation logistics. That means I'm in management here at the trucking company."

Since 1968, he's also edited a scientific worm journal called *Megadrilogica*. I keep a stack of back issues on my desk, next to a pot of earth that sometimes contains a worm I've brought in from the garden for a short stay. Each issue of the journal is printed on eleven-by-seventeen-inch paper that, when folded in half and stapled, makes for a neat letter-sized journal that is comforting in the absolute black-and-white certainty with which it elucidates its slippery and evasive subject. There are no illustrations, only a few maps and tables, and long lists of academic citations in the back of each issue. The articles have titles like "Bacteriology of Laying Hen's Manure, Composting, and *Eisenia Foetida* (Oligochaeta: Lumbricidae)," "Note on an Indian Earthworm With Two Tails," and the wistful "Farewell to North American Megadriles." Reynolds's own articles tend to have straightforward titles like "The Earthworms of New Brunswick," "The Earthworms of South Carolina," and "Primeros Datos de Lombrices de Tierra (Oligochaeta) de la Isla de San Andrés, Colombia" (roughly translated: "The Earthworms of San Andrés Island, Colombia").

In spite of—or perhaps because of—his various careers, Reynolds has always found a way to get out and collect worms. Lately he's discovered that truck driving and family vacations provide the best opportunities. "When I was driving a truck, I could pull into any rest stop, turn over a log, and find some worms. Now I'm in management, but some of the truckers pick up worms for me. I send them out with specimen collection kits and they stop off at rest stops and gather worms for my research.

"And this year we're off to Panama for vacation. I can get the best worms at tourist resorts. They bring in all these plants from the jungle to plant around the hotel, and there's always worms in the roots. That's the global distribution of earthworms for you, right there in those ball and burlap bags. I just go around and talk to all the gardeners and get them to bring worms to me. They'll tell me where they found them, what plants were growing nearby, everything."

John Reynolds has traveled around the world, gathering and classifying specimens. His collection of one hundred thousand earthworms recently went to the Canadian Museum of Nature. The collection had simply grown too large to curate properly, he told me, and it was something of a relief to let it go to an institution that could make better use of it.

More than any other earthworm scientist I've met, Reynolds laments the lack of popular support for worm research. He once wrote, "I have long advocated the necessity for our scientific information to be more accessible to the general non-scientific community. We are a small group of scientists working with limited finances and materials. We have suffered from not following the examples of entomologists and ornithologists. These disciplines have advanced more rapidly because of the contributions of 'amateurs' and general collectors." Still, he faces an uphill battle in his quest to rally the public around the cause of earthworm science and preservation. "I do radio, television, public lectures. People ask me, why bother cataloging earthworms? Well, why catalog anything? It's how we learn about the world we live in.

Besides, some of these worms are going extinct. How do you know what you're losing if you don't know what you have?"

Once an earthworm is discovered, it has to be classified and named. Worms are grouped under the same branch of the animal kingdom's family tree, phylum Annelida, as leeches and aquatic worms, pointing to their early connection with water-dwelling organisms five hundred million years ago.

Within the phylum Annelida, earthworms are organized under the class Oligochaeta, and beyond that, taxonomists disagree. Recent classifications have identified two orders, several suborders and superfamilies, 23 families, 739 genera, and over 4,500 species. Here is an example of earthworm taxonomy for *Lumbricus terrestris,* which was identified by Linnaeus in 1758 (and was the only one he described):

| | |
|---|---|
| Kingdom: | Animalia |
| Phylum: | Annelida |
| Class: | Oligochaeta |
| Order: | Haplotaxida |
| Family: | Lumbricidae |
| Genus: | Lumbricus |
| Species: | Terrestris |

The constant shifting and reclassifying of earthworms is frustrating to all but the most dedicated taxonomist. When I complained to Sam James about all the contradictions in the literature, he said in a matter-of-fact voice, "This is routine with emerging science. We're fixing mistakes. In the next century, we'll start to see the complete family tree."

Oligochaetologists today are studying questions that, for many insects and animals, were answered in the last century. That's not to say that a great deal of progress has not been made: since Darwin's day, scientists have managed to quantify the extent to which earthworms not only plough the earth, but transform the organic life of the soil. But it could be years before the complex relationships between plants and soil-dwelling creatures are fully understood. Meanwhile, taxonomists like Sam James are still identifying and describing new species. When I saw the wooden case in Sam's laboratory filled with vials of preserved worms, each shelf in the case marked by the country in which the worm was found, I told him that his collection more closely resembled that of a nineteenth-century naturalist than that of a twenty-first-century researcher.

"This is the sort of collection that I picture the young Darwin putting together in the 1830s after his trip around the world on the *Beagle*," I said. "It looks so old-fashioned. How many scientists today get to fill their labs with new species like this?"

"Not many," he said. "You're right. In some ways, we are back in the nineteenth century."

It must be both a joy and a frustration to work on the frontiers of science. Sam James, like most of the oligochaetologists I met, is a passionate advocate for the subjects he studies, genuinely unable to understand why everyone isn't as fascinated as he is with earthworms. It is reasonable for such a person to look at the many societies established for the study and appreciation of birds or butterflies or at the laws in place to protect dolphins and starfish, and the lengths to which people will go to attract ladybugs or honeybees to their gardens, and wonder why no comparable efforts have been made on behalf of earthworms. Are we so hierarchical that we can't respect a creature that lives beneath our feet? Are we so focused on image, on appearance, that we can only love the prettiest inhabitants of the garden—a swallowtail butterfly, a fat bumblebee—and neglect the slimy but hardworking earthworm? Perhaps it is because we associate them with death and decay, while bees bring to mind sunflowers and sweet honey and the mild sexual buzz of a flower bed being pollinated.

Still, earthworm scientists work on, alone, undeterred by the lack of popular support. Their discoveries—the unusual and extraordinary worms they collect—might someday be stored at the Smithsonian or the American Museum of Natural History, but they may not be exhibited for years. You can get a map that shows the distribution of red fire ants across the southern United States, one that shows where redwood forests used to grow worldwide and where they remain today, even a map showing the location of the endangered Oahu tree snail. But you can't get a reliable map of the distribution of earthworm species worldwide, precisely because worms live underground and must be dug up, one at a time, in order to be identified and counted. Once they've been discovered, it could be months or years before a researcher has time to properly name and catalog the species.

I once asked Sam if he knew anything about earthworms in the Amazon basin. He said he could think of a few, including some giant worms over two feet long. When I asked him for a Latin name, he said, "I don't know. I haven't named them yet." If he doesn't name them, I wondered, who will?

# Intestines of the Soil

What would our lives be like if we took earthworms
seriously, took the ground under our feet rather than the skies
high above our heads, as the place to look, as well, eventually,
as the place to be? It is as though we have been pointed
in the wrong direction.

—ADAM PHILLIPS, *Darwin's Worms*, 1999

"ARCHAEOLOGISTS ARE PROBABLY not aware how
much they owe to worms for the preservation of many ancient
objects. Coins, gold ornaments, stone implements, &c., if
dropped on the surface of the ground, will infallibly be buried by
the castings of worms in a few years, and will thus be safely preserved, until the land at some future time is turned up."

Darwin made this proclamation in the beginning of the
fourth chapter of *The Formation of Vegetable Mould*, entitled
"The Part Which Worms Have Played in the Burial of Ancient
Buildings." He went on to describe the excursions he or his son
William took to excavation sites around England, including a
farm in Surrey where Roman ruins were found, an abbey in
Hampshire destroyed by Henry VIII, and the ruins of a Roman
villa in Gloucestershire. He reported that worms had burrowed
into the old stone walls, undermined foundations, and generally

deposited a layer of castings that permitted grass and other plants to grow. After examining the sites of several ancient ruins, he concluded that the actions of earthworms "would ultimately conceal the whole beneath fine earth."

In some ways, Darwin thought of worms as historians, covering the remains of one civilization and preparing the earth for the next. But earthworms can hardly be considered sneaky in their concealment; in fact, anyone who has ever watched a worm knows that it goes about its work in the most matter-of-fact manner. It is only carrying out the natural order of things, folding the ruins of a city, a farm, or a society into the lower strata of the earth. When our civilizations end, and when we as individuals die, we don't ascend, not physically—we descend. And the earth rises up to meet us.

I read about Darwin's visit to Stonehenge, where he saw firsthand that earthworms were drawing those ancient rocks down to their underground world, and I shared in his sense of wonder over the continual job of burial that they performed. Earthworms toil incessantly to carry us down to the depths of the earth with them, and ultimately, our efforts to resist are futile. Darwin must have taken some comfort in this; he never feared death, and he appreciated the natural order that this arrangement with earthworms suggested. In a way, looking down at those buried Roman ruins must have been like getting a glimpse of a kind of afterlife.

But even Darwin knew little about the civilization that existed down in the earth. The clumsy artifacts he uncovered would have held little interest to him if he had instead been given one glimpse into a microscope powerful enough to illuminate all the invisible creatures that inhabit the soil alongside earthworms. A new world would have been laid bare to him. Like a shipwreck diver distracted from his task by the aquatic world around him, Darwin would have soon forgotten about his ancient ruins.

Eternity can be found in the minuscule, in the place where earthworms, along with billions of unseen soil-dwelling microorganisms, engage in a complex and little-understood dance with the tangle of plant roots that make up their gardens, their cities.

A BACKYARD GARDEN is a miniature ecosystem, one that requires constant tinkering. Gardeners are always assessing whether their plots of land could use more alfalfa meal, less water, another layer of manure or mulch. Even the insect population can be modified, managed. Ladybugs can be purchased, lacewing eggs hatched, honeybees lured in, all in the name of keeping the garden in balance. Now even microscopic creatures can be ordered by mail and introduced to combat some pest, seen or unseen, wreaking havoc in the flower beds. It wasn't until I met up with some of those mail-order microscopics that I began to get a glimpse of the vast unseen world in which earthworms are such a vital force.

When I moved to Eureka last year, I knew things would be different in the garden. The temperature here is always five or ten degrees cooler than it is in Santa Cruz. Fog hangs over Humboldt Bay most of the year, making a full day of sunshine a rare event, one that most homeowners celebrate by getting out to scrape the moss off the north side of the house. The seasons change slowly. Some would say that the only real difference between summer and winter—apart from a few blessed degrees of warmth—is that it stops raining in the summer.

This gave me plenty to think about when I first planned my new garden. The sage plants I relied on for year-round bloom would have trouble getting established here. Heat-loving vegetables like tomatoes and peppers would surely struggle in the cool, damp summer, although I would plant some anyway in the faint hope that my backyard would be warm enough for them. The angel's trumpet, with its enormous, bell-shaped flowers, would certainly succumb to frost damage in winter. Only the lilac, which prefers a chilly winter, would do well here. I'd have to learn to appreciate the plants that would grow in this climate: rhododendron and azalea in the flower garden, cabbages and potatoes in the vegetable garden. In the middle of winter, when the moving van arrived packed full of plants from my Santa Cruz garden, I did what I could. I put them in the ground where I thought they'd do best, but it would be a year or more before they got established and the garden began to have a structure. I spent that first year nurturing tiny transplants along, heaping compost around their roots and pulling weeds. Soon the garden and I were in our second spring together, and it was showing me what it could do. The lavender hedge was filling in, and the daisies and coreopsis showed signs of a long blooming season ahead.

One thing I wasn't prepared for was the insect population. I saw snails and slugs everywhere—on the buds of my daffodils, chewing young green pea shoots. There were aphids, but I was also pleased to find an abundant population of ladybugs, and they would keep the aphids in check. The real trouble involved the new pests in this garden, pests I'd never done battle with before. Furry black caterpillars massed on the underside of my artichoke leaves, eating holes until the leaves looked like fine lace. Tiny green caterpillars were making a meal out of my cabbages. Worst of all, some unseen enemy was sawing off seedlings as they emerged from the soil.

The first two pests were easy enough to control. The black caterpillars—tent caterpillars, I learned they were called—could either be picked off and dumped into a pail of soapy water or sprayed with an organic treatment that causes them to stop eating and slowly starve to death. The cabbage loopers, moth caterpillars that feed on cole crops, could be kept in check with the same spray, or I could encourage their natural predator, a parasitic wasp, to set up camp. Since parasitic wasps happen to like many of the same flowers I do—yarrow, tansy, alyssum—this was an easy concession to make. That left one more to conquer: the one cutting down my seedlings from the root.

I'd been paying a lot more attention to the soil now that I was keeping tabs on my earthworm population. I had a general idea of the kind of creature I was looking for—a grublike caterpillar called a cutworm that lived near the surface of the soil. I watched for it any time I was out digging in the dirt. Sure enough, before long I spotted one. It matched the description in my pest control book: "a fat, greasy, grey or dull brown caterpillar with a shiny head; found in the soil." This one was curled in a tight circle just a few inches below the surface. Over time I would find that cutworms usually stay in that position during the day and venture out after dark to do their damage.

I flipped the grub onto my shovel and flung it into the street. This is how I usually handle snails, too: I toss them into the road and let cars drive over them so I don't have to squash them my- self. But it's one thing to walk around the garden pulling snails off leaves and tossing them into the road; it's another thing en- tirely to dig around in the dirt for cutworms. I could dig up half the garden looking for them.

It turns out that the best control for cutworms is a particular type of parasitic nematode, a microscopic creature that lives in the soil. It enters a cutworm's body, releases bacteria that kills it, then feeds off the dead cutworm and lays eggs. I'd never added nematodes to my own soil, but I knew that garden supply com- panies sold them for any number of pest infestations: one worked against fleas in lawns, another worked against mole crickets, one attacked cutworms and squash vine borers, and one even at- tacked other nematodes like the destructive root-knot nematode that often plagued carrots.

I chose the nematode that was most effective against cut- worms. I started small, ordering just five million. They arrived inside a slim sponge not much larger than a credit card. The sponge was sealed inside a plastic bag. A yellowish liquid oozed out of it. These, I took it, were the nematodes, immersed in some kind of solution. The whole package was smaller than a letter- sized envelope. All I had to do was drop the sponge in a bucket of water, wring it out, and spray the water around my garden.

How could five million creatures be alive in that sponge? How could something so small kill a cutworm, an inch-long grub? How would they even find them in the soil? I felt a little silly, splashing my bucket of water around the garden. Without a microscope, without any real scientific training, I was operating mostly on faith. I couldn't see the nematodes, but I had to believe they were there, and I had to believe they'd find the cutworms somehow.

Because most nematodes are microscopic, it is easy to overlook their role in the soil. But Nathan Cobb, a pioneer in the science of nematodes, wrote, "If all the matter in the universe except nematodes were swept away, our world would still be dimly recognizable, and if, as disembodied spirits, we could then investigate it, we should find its mountains, hills, vales, rivers, lakes, and oceans represented by a film of nematodes." It's a creepy thought, all those invisible organisms covering us like a thin film. Even worse, one study reported about ninety thousand nematodes living in a single rotten apple. The soil is quite literally crawling with them: several thousand can be found in a half-cup of soil. The problem is that not all of them are good; some suppress plant diseases and others cause diseases.

Even though there are nematodes that can harm plants, their presence is generally considered a sign of good soil health. They are decomposers, feeding on bacteria and fungi, and sometimes parasitizing larger soil-dwelling creatures like the cutworms I was trying to get rid of. They're also a major food source for earthworms, and certain species even rely on earthworms as a "reservoir host," a vessel to shelter them during part of their life span. (I once saw a collection of tiny, barely visible nematodes, all gathered from the intestines of earthworms during dissection. About a dozen of them were in a vial together, and when they were held up to the light, looked like nothing more than dustlike particles drifting in a clear solution.)

Nematodes, along with the bacteria, fungi, and protozoa that inhabit the soil, are the unseen companions of the earthworm, serving as a food source, a collaborator, or — at times — a passenger in the earthworm's gut, traveling distances and even finding a permanent home in the worm's nutrient-rich intestine. This is the earthworm's powerful secret, one that even Darwin didn't fully grasp: the earthworm, far from being one of the smallest and weakest creatures, is actually one of the largest beings in its world, its underground society. In that place, it is an elephant, a whale — a giant.

SOIL BIOLOGISTS DESCRIBE what happens among the inhabitants of the soil in terms of a "food web," as opposed to a "food chain," for good reason: a chart of the complex relationships between earthworms and the other soil-dwelling insects resembles a spider's web more than it does a hierarchical stepladder. The other inhabitants, apart from earthworms, are arthropods: insects such as mites, ants, spiders, millipedes, scorpions, beetles, sowbugs, and springtails. (Not all are visible to the naked eye: some, like certain mites and springtails, are microscopic.) These are often called "shredders" for their role in breaking down fallen leaves, bits of bark, and other organic matter. One soil scientist wrote that without arthropods, a bacterium in a pile of leaves would be like "a person in a pantry without a can-opener." Like earthworms, arthropods break down the soil, build burrows through it, and add nutrients in the form of their castings. They outnumber earthworms by roughly four hundred to one: a square yard of garden soil might contain a few hundred earthworms, but it could contain one hundred thousand or more arthropods.

EARTHWORMS, THEN, ARE FAR from alone. Nowhere is this more evident than in my bin, where worms share space with a number of other creatures. I can't say for certain how they got in the bin—because it sits outside, I suppose they just wandered in—but over time, I've found that they're mostly harmless and may actually help worms do their jobs. So now I don't mind having them around.

Pill bugs were one of the first creatures to crawl into the bin. Also called sow bugs, these are the little roly-poly crustaceans that carry hard grey armor on their backs and roll into a ball when provoked. They feed on decaying matter and are drawn to the damp conditions of the worm bin. They don't harm the worms, and they're not bad to have in the garden either, so I let them stay in the bin. When it's time to harvest a tray of castings, I probably release a third of the sow bug population back into the garden, where they might, for all I know, make their way back to the worm bin eventually.

I also have a healthy population of pot worms. The proper name for them is enchytraeid worms; they are a distant cousin of earthworms and are sometimes mistaken for baby worms. But baby worms are faintly reddish and translucent; pot worms are solid white. They favor the same food source as earthworms, and their castings also become food for other microorganisms in the soil. I honestly don't know how they made their way into the bin—I can't imagine that they would come out of the soil and climb the plastic walls just to join the earthworms in their feast, but there they are. They work alongside the worms, often starting on food before the worms do. Perhaps they even help to break down the larger pieces of food—a chunk of cauliflower, a heel of stale bread—so the worms can go to work on it, too. That's just my theory, but even if they don't help, they certainly don't hurt. I leave them alone, also.

For a while, I got so interested in the other inhabitants of the worm bin that I tried to add one into the mix. A local worm farmer showed me his bins, and I was surprised to see that the castings were covered in tiny white dots. It looked like someone had sprinkled salt or sugar on them, except that when I bent down to touch a few of the white spots, they jumped away. These were springtails, the same tiny white creatures that are sometimes found under a pile of leaves or at the base of a rotten log. They also feed on decaying plant matter, and they can be a vital part of a healthy soil community. I didn't know why they had never appeared in my worm bin, but suddenly I wanted some. The farmer let me scoop out a shovelful of castings, which held more tiny springtails than I could count. At home, I added them to my own worms, but something in my bin must not favor the development of springtails, because when I went back and looked for them a few days later, I didn't see any.

WHAT IS HAPPENING in my worm bin is really a microcosm of what's going on in the soil. Add up the number of earthworms and other soil-dwelling creatures like mites, springtails, ants, and spiders, and there may well be more living things in one of my four-by-four vegetable beds than there are humans in all of rural Humboldt County where I live. Include the nematodes, and the population of one of those vegetable beds starts to rival that of the state of California. An earthworm could begin to feel a little crowded in such an environment. But it doesn't end there: bacteria, fungi, and protozoa inhabit the soil—or the damp, dark confines of a worm bin—in far greater numbers. One teaspoon of soil could hold a billion bacteria, for example. It is here—at this microscopic level—that scientists are only just beginning to appreciate the critical role that earthworms play in the web of life that exists underground.

Picture a pinch of soil under a microscope. Large, irregularly shaped boulders—the clay and sand particles—stand surprisingly far apart from one another. The space between those particles makes up about half the volume of good, loose soil. In addition to allowing water and air to penetrate, the spaces give living organisms room to crawl around. Those creatures, along with the rotting organic matter they feed upon, account for only ten percent of the overall soil volume and show up under the microscope as tiny spots moving between the soil particles like black ants crawling around a rock. If an earthworm were to come across this scene, it could devour the whole community— soil particles, nematodes, bacteria, protozoa, and fungi—in one swallow.

To understand how earthworms interact with their microscopic counterparts, it is probably best to start with the single-celled bacteria that inhabit the soil by the billions. Bacteria help compost piles to decompose, creating the right environment for epigeic earthworms like *Eisenia fetida*. These worms won't eat until bacteria have come in to start breaking down their food source. In fact, as long as the worms in my own bin have enough to eat, I've noticed that they will leave one of their favorite foods, a banana skin, alone for several days until it is more decomposed.

Some worm farmers recommend speeding up this process by inoculating new worm compost systems with bacteria-rich castings from another worm bin, or even by letting the food decompose in an ordinary compost pile before feeding it to them. If I had a much larger quantity of food to feed to my worms, I might do this myself. As it is, I've got only a handful of food scraps to give them each day; it is easier to scrape them right off the cutting board into the bin. They will decompose soon enough, and there's always something else in the bin for the worms to eat that is farther along in the process.

Bacteria also contribute indirectly to the health of earthworms by helping plant life to flourish: some are responsible for converting ammonium into nitrate, a form of nitrogen that is easier for plants to use, others help fix nitrogen in the roots of plants, and still others help break down carbon, sulfur, hydrogen, and other compounds. In addition to transforming these nutrients into a form that is easier for plants to access, bacteria can help stabilize them, making them available in the soil for a longer period. A particular type of bacteria is even responsible for the fresh, earthy smell of soil. And although they are crucial to the health of soil and provide an abundant food source for earthworms, not all bacteria are beneficial. Some cause plant diseases. For example, the disease that causes squash and cucumber vines to wilt and collapse is caused by a particular soil bacteria.

What does it mean to a bacteria population, then, if earthworms are present in great numbers? One study showed over fifty different species of bacteria living in the gut of one nightcrawler. The intestinal mucus of a nightcrawler is an excellent food source for bacteria; they can thrive and reproduce inside the body of a worm, until far more bacteria emerge from the end of a worm than entered it in the first place. Still, an earthworm does not have an equally beneficial effect on every species of bacteria. Some are killed as they move through its intestinal tract, while others grow so rapidly that they can become the majority bacteria in a soil where they otherwise would have existed only in small quantities. Also, different species of earthworms may have unique relationships with certain bacteria. There's been very little research done on this point, but when I asked oligochaetologists about it, all agreed that it was an interesting notion. "We don't know for certain which species of earthworms might help a particular bacteria to flourish and to what extent," one researcher told me. "But there are plant diseases that are caused by specific bacteria or fungi. If we found out that earthworms helped to spread those microorganisms, or helped to destroy them by encouraging their natural predator bacteria, that could be very useful information."

IF BACTERIA CAN BE pictured as teeming black ants under the microscope, imagine fungi as gossamer spider webs. These organisms form long threads called hyphae that stretch between plant roots. Some form into even larger masses called mycelium that can span an entire backyard. Still others, called mycorrhizal fungi, cluster around plant roots and help bring nutrients and moisture to the root zone in exchange for the carbohydrates their host plant provides. Some farmers and gardeners who believe that these fragile, microscopic networks are too easily disrupted by deep ploughing, advocate churning the soil as little as possible. After all, fungi help decompose nutrients in the soil and keep them in place so that plants can use them.

But like bacteria, some fungi are harmful to plants. The verticillium wilt that plagues tomatoes each year is caused by a fungus, as are a variety of root rots that strike vegetable gardens. Once again, the best way to fight a fungus is with another one: the beneficial *Gliocladium virens* can prevent destructive fungi from getting established in soil. I have ordered it for my own garden in the last few years. Like the nematodes, this fungus comes in a surprisingly small package: one small foil pouch contains enough fungi, in the form of an ordinary-looking white powder, to inoculate most of my vegetable garden.

This spring my shipment arrived just in time; a week later and I would have been ready to put tomato plants in the ground. I tore open the package and walked through the vegetable garden, a shovel in one hand and the packet of beneficial fungi in the other. I stopped at each bed, turned the soil over, and sprinkled a little *Gliocladium virens* into the ground. It occurred to me that by distributing this fungus through my soil, I was changing the microscopic population of the earth. I was doing on a large scale what earthworms do on a small, but far more effective, scale.

There are all kinds of ways to tinker with the invisible life of the soil. Consider the way that researchers have used the earthworm in the fight against the take-all fungus *Gaeumannomyces graminis.* Take-all got its name from its ability to wipe out an entire field of wheat or barley in one season. The bacteria called *Pseudomonas corrugata* is effective against take-all, but researchers found that it was slow to move through the soil and attack the fungus. However, if it was mixed into sheep dung and then introduced to the soil, earthworms would eat the sheep dung and carry the bacteria through the soil, helping to move it more quickly to the roots of plants, where it could do its job.

Another wheat fungus, *Rhizoctonia solani* or bare-patch disease, decreases as the presence of particular earthworms increases, presumably because this fungus can't survive the passage through a worm's gut. But killing off a harmful fungus isn't the only way earthworms can eliminate plant diseases: sometimes they help by carrying the fungus away from its host plant. One study showed that trees infected with apple scab benefited from large populations of nightcrawlers: the worms consumed the fallen leaves around the trees and carried the fungus on those leaves away with them in their guts, helping to disperse the colony of fungi that would have infected the tree anew in the spring.

What makes this situation so complex is that earthworms may help a particular fungus survive while killing off another one. Some of the most beneficial fungi—those most responsible for plant growth and nitrogen fixing—increase dramatically when they're present, presumably because those fungi flourish in the nutrient-rich intestine of an earthworm. But other, harmful fungi have also been shown to thrive in the presence of earthworms. Dwarf bunt, a winter wheat disease, is one such fungus that can be spread by them. The dreaded tomato plant fungus, fusarium, might also be spread by an active worm population. Earthworms could have no particular effect on one harmful fungi, but might eliminate or help to spread that fungi's main predator, thus tipping the scales in one direction or another. It seems unlikely that the risk of earthworms spreading diseases could ever outweigh their benefits. What we know for certain is that the earthworm is a powerful influence in the soil.

BACTERIA, FUNGI, AND nematodes are perhaps best known to gardeners like me because of the diseases they cause or combat. But from an earthworm's perspective, protozoa may be the most important organisms in the soil. These multicelled creatures are quite a bit larger than bacteria, although still microscopic and still smaller, on average, than nematodes. Protozoa feed mostly on bacteria, competing with nematodes for this particular food source. A soil that is abundant in nematodes may have fewer protozoa, or vice versa, and even today scientists don't know what this might mean to plant life. What is known for certain is that protozoa make up an important part of most earthworms' diets. In fact, one study showed that *Eisenia fetida* could not reach sexual maturity unless protozoa were present in the soil. They thrive in compost piles, where they consume bacteria and, because bacteria contain more nitrogen than the protozoa need, they release excess nitrogen back to the soil. Like bacteria and fungi, many protozoa survive the trip through the earthworm's gut and are found in large quantities in earthworm castings. Not all protozoa are equally nutritious to an earthworm: one scientist fed *Eisenia fetida* a restricted diet of two particular protozoa, and the worms died within a few days.

That brings up one of the great curiosities of earthworm life: in spite of all the microscopic creatures living alongside earthworms and inhabiting their guts, they seem to have few enemies in the soil. There have been a few stories of creatures that move into an earthworm's body as a parasite and eventually destroy it: in Illinois, someone found a nightcrawler that was near death after a particular kind of nematode had taken it over. The nematodes caused a bacterial infection that killed the worm, then thousands of nematodes emerged from the dead worm's body. (I saw a photograph of this event. I wish I hadn't. The nematodes, at great magnification, themselves looked tiny glistening white worms. They swarmed over their host, covering it so completely that I could not see the nightcrawler at all. It was a horrifying sight.)

But stories like this one are rare. More likely than not, a worm will not be harmed by the organisms passing through its gut and living alongside it in the dirt. Worms have more to fear from birds, moles, mice, and rats. They seem to suffer from very few parasitic or bacterial illnesses caused by the tiny creatures in their world. It is just another way in which worms seem to have a powerful ability to survive. An infection of streptococcus bacteria could, if left untreated, make me seriously ill. The sturdy constitution of the worm makes humans seem weak in comparison.

This once led me to call Sam James and ask him a silly question: how do worms die? If a worm manages to escape the robin's beak, and if it is generally unmolested by parasites or other illnesses, what would kill it? And if they die of old age, what does that mean, exactly? In people, the symptoms of old age are obvious. A person might become frail and weak, their bones brittle, their skin fragile. One might die in old age of heart failure or cancer or a stroke. But what are the symptoms of old age in an earthworm? And why have I never found a dead worm in the soil, or, for that matter, in my compost bin?

Sam said that part of the reason I had never seen one is that they decompose so quickly. He told me that he once watched a dead worm in an underground root facility. "It lasted all of twenty-four hours before it totally vanished," he said. "And I found a very nearly dead worm under a log once. Those are the only ones I have ever found in nature."

As for how and why they might die of old age, he didn't know, and I can only assume that if he doesn't know, no one knows. There are no features that allow a person to tell a worm's age, he explained. As long as it is an adult, a young worm looks the same as an old worm. Perhaps it makes sense that a creature that doesn't get ill and has few enemies among its neighbors would also live agelessly and die without explanation or cause—would simply vanish without a trace.

WORMS PICK UP all kinds of odd organisms and objects as they eat their way through the dirt. They'll swallow a tapeworm whole, for instance, and they can even consume a parasitic egg, burrowing deeper into the soil, where the egg is excreted in dark, safe surroundings that protect it from harm until it hatches. That makes farmers wonder if they could be a vector for transmission of parasites on the farm. And in the same way that worms can spread bacteria and other microorganisms, they have also been blamed for the spread of weed seed. A tiny, smooth, hard-shelled seed can pass quite safely through their guts. Researchers have observed worms choosing particular seeds over others and have suggested that in the right environment, they could act as powerful agents to spread the seeds of those plants they prefer.

Here's an example of how intricate the relationships between worms and other microbes can be. A researcher once fed the bacterium *Pseudomonas corrugata* (which causes pith necrosis disease in tomatoes, but may also combat certain other plant diseases such as ring rot disease in potatoes) to four species of earthworms. One anecic worm, *Aporrectodea longa,* had ten times more *Pseudomonas corrugata* in its castings than did the other three worms in the study. If a farmer knew that *Aporrectodea longa* existed in the soil, he or she might be encouraged to know which worms could help spread bacteria that might prevent ring rot disease in potatoes. But if that farmer decided to plant tomatoes, those same worms spreading the same bacteria could put the tomato crop at risk for pith necrosis disease.

What's a farmer supposed to do with this information? What could I do, in my own garden, to shift the subterranean population so that it favors the fruits and flowers I like to grow and keeps diseases and pests at bay? When I think about it like this, I realize that I have started to garden upside-down. I've begun to tend the soil the way I'd tend a farm, treating the bacteria, fungi, protozoa, and—of course—the earthworms, like a kind of livestock. When I add a shovelful of worm castings to the bottom of a planting hole where a new plant is about to go, I know that I am inoculating the earth with a concentrated population of invisible beings that I believe will shift the balance in the soil, creating a powerful community that will live at the feet of my new plant and perhaps even attract earthworms to the spot, where they will work to maintain and encourage the roots.

This is crude plant science, to be sure. But farmers and gardeners are increasingly turning to biological solutions for plant diseases and insect infestations, from ladybugs and parasitic wasps to nematodes and fungi. It does not seem so far-fetched, then, that farmers or gardeners could inoculate their soil with a particular earthworm that was known to do the most good—or, in the case of bacteria that cause plant diseases—the most harm. We know now that earthworms do more than plough the earth. They are at times the movers, incubators, or destroyers of the invisible denizens of the soil. But the study of earthworms is still in its infancy where soil microorganisms are concerned. It is far too early in the game for a scientist to recommend a particular earthworm as a biological solution to a farmer's problem. Still, it is becoming clear that earthworms are, as one researcher put it, "a keystone species." When an earthworm is introduced into a forest, a farm, or a backyard garden, it can bring about profound changes at the microscopic level, changes that can transform plant life aboveground.

# As They Can neither See nor Hear

Have you no beginning and end? Which heart
is the real one? Which eye the seer? Why
is it in the infinite plan that you would
be severed and rise from the dead like a gargoyle
with two heads?

—ANNE SEXTON, "Earthworm," *45 Mercy Street,* 1976

"WANT TO SEE AN anatomically correct toy earthworm?" Sam James asked me when I walked into his laboratory.
I flew to Iowa to meet Sam and to see the rare specimens he's
collected on trips around the globe. I came prepared to see every
kind of worm except the plastic variety.

He didn't wait for me to answer. "Look at this," he said, laying a long plastic worm on the counter in front of him. "If you
count the segments"— here he ran a pair of tweezers along the
length of the toy worm, tapping each segment as he went—
"you'll see that it has exactly the same number of segments between the anterior end and the clitellum as the worm it was
modeled after. And look at this!"

He turned the worm over. Underneath the clitellum were the
words "Made in China" in raised letters. But that wasn't what
Sam was excited about. He was pointing to a pair of raised white

dots just outside the clitellum. If this was my toy worm, I would have assumed that these dots were just a defect in the manufacturing process.

"Sexual pores!" Sam exclaimed. "They got them right! Everything about this worm is dead accurate. Somebody decided to make a toy worm, and they got an actual worm and made a perfect copy of it. Amazing."

It's easy to see why this level of precision would be so exciting to an oligochaetologist. Although worms are fairly simple creatures, they are not as featureless as most people think. When Sam pulled a preserved worm out of a jar of formaldehyde and laid it under the microscope, the nearly invisible details of its anatomy leapt into focus.

"Look there," he said, pointing at the nubby bristles that ring the worm's body. "The setae. Most worms have eight or twelve in pairs around each segment. On this specimen, they go all the way around the worm."

"And the sexual pores . . . here . . . ," he said, indicating several pair of male pores and one set of female pores. The location of these pores, the number of segments, the location of the setae, and the length and color of the worm give a taxonomist like Sam James his first clues when he's trying to identify a worm. Beyond that, dissection is the only way to know for sure.

He pulled out a color drawing of an earthworm's internal organs that he uses to teach students. "One to five pairs of hearts, of course," he says matter-of-factly. "The position of the last pair of hearts—which segment they're in—can give you a clue about the species of worm you're looking at. Also, you want to check to see whether the hearts are connected to the dorsal blood vessel or not." (Worms have just a few major blood vessels running the length of their body, and in one of the more translucent species, you can see the vessel running along its back.)

"Let's see, what else . . . ovaries are always in segment thirteen. . . ."

"Always?" I asked.

"Well, sometimes they're in segments twelve and thirteen. Oh, and I guess there are a few other exceptions. But that's usually where you'll find them. And they've got a gizzard to help grind up food. It's always around segment five, six, seven, eight. The location of the gizzard along the intestinal tract can give you a clue about the species. Then we look at the calciferous glands along the esophagus—those are used in digestion—and the placement of these nephridia, which are excretory organs. There could be several of them."

I looked over the drawing. "No lungs," I said.

"Right. They breathe through their skin."

"Do they sleep?"

He smiled. I imagined we were thinking the same thing. How would you tell if a worm is asleep, short of hooking up electrodes to it? They don't snore.

"No," he said. "They don't seem to sleep. They have periods of inactivity, but that's not sleep as we understand it."

"And they don't have much of a brain."

"Well," he said, quick to defend the creature he's spent a lifetime studying. "The cerebral ganglion is here, around the third segment. It's a primitive brain. You've read Darwin. They're smart enough to figure out how to pull a paper triangle into a burrow."

I WALKED AROUND Sam's laboratory, peering at the pickled worms in jars. During the morning I spent there, I didn't see a single live worm. As a taxonomist, he's mostly concerned with identification and dissection. Even if he wanted to study their behavior, he'd find worms impossible to track in the wild and difficult to observe in the laboratory. They don't perform well under glass: they avoid light and don't take to artificial conditions. In fact, complete life cycle studies have been performed on fewer than a dozen species, precisely because it is nearly impossible to study an earthworm in its natural habitat. A marine biologist can drop a submarine into the middle of the ocean; an astronaut can drift among the stars in a space capsule. But the middle of the earth is unexplored precisely because it is so difficult to get there. We spend our lives on the surface of this planet, but we've devised no good way to explore the center of the earth, to observe its inner workings. We've managed to make small inroads: geologists drill into the soil and pull out samples, miners walk through underground tunnels with flashlights, and plant scientists sit in a root observation facility called a rhizotron—a glass-encased underground room in which one can watch roots grow or wait for an earthworm to slither past—but in the end, we are left with precious little information. The ocean and the heavens are transparent; this simple fact renders them more accessible to us than the earth will ever be.

It's not easy, then, to know how earthworms live out their lives, because it is so difficult to watch them do it. Darwin made some good guesses, and scientists working in the decades after his book was published have managed to paint a more complete picture of earthworm life underground. One major accomplishment has been to describe their sexual behavior. After all, worms are no more inclined than any other creature to perform the act in a laboratory's artificial setting, and their natural preference is to submerge themselves in the cool, dark earth, away from a scientist's prying eyes. Still, researchers have been persistent, scraping away the soil in a terrarium, forcing a pair of copulating worms against the glass walls of a jar, and they've seen enough that they've managed to gather a few elementary facts.

When a worm is looking for a sexual partner, its primary criterion is length. A pair of worms mate by slithering alongside one another, belly to belly, head to tail. Because worms are hermaphrodites—they have both sets of sexual organs—they must arrange their bodies in such a way that the male organs of one line up with the female organs of the other. These sexual organs— really just tiny pores—tend to be located around the twelfth segment, between the worm's head and its clitellum. Once the worms are in position, the male organs release sperm, which is taken in by the female organs. For some worms, including the nightcrawler *Lumbricus terrestris,* the pores aren't designed to line up precisely, and seminal fluid has to travel along a groove that is formed when the worm contracts the longitudinal muscles in its body, forcing drops of fluid towards their final destination one segment at a time. Many species have several pairs of female apertures, and copulating worms will adjust their position a few times so that sperm is deposited in each one. (Experts will tell you to be highly suspicious of claims of "superworm" hybrids or crosses between two species: the logistics alone make it highly unlikely that worms of different species could copulate successfully.)

The act can take a few hours to complete and during that time the worms are oblivious to their surroundings. Even Darwin took note of this, writing, "Their sexual passion is strong enough to overcome for a time their dread of light." He took this as a sign of their intelligence. The simple fact that they could be so focused on one activity that they would not notice anything else led him to conclude that they must have a mind of some kind. If they were merely beings that reacted to stimuli by instinct, he reasoned, they would dodge the light no matter what else they were doing.

Worms have developed several strategies to keep themselves anchored to each other during copulation. I can understand why this is necessary: in my own worm bin, thousands of worms are in motion at once, all sliding over and under one another in search of food. It would be easy, in that damp environment, with no hands or feet, to get knocked off course. Some worms grip each other with their setae, using the shorter and sturdier bristles near the sexual organs. They also excrete a great deal of sticky fluid that helps keep them anchored to each other. This fluid covers the length of each worm's body, from the reproductive organs to the clitellum, forming a sort of tube around the body that sticks to a similar tube on the other worm.

The worms separate from each other after a few hours, but their eggs are not fertilized right away. Instead, the egg and sperm are held together under a heavy mucus secretion produced in the clitellum. That secretion hardens eventually—this process can take a few days—and the worm wriggles backwards, sliding out of the mucus shell, which now contains egg and sperm. The ends seal as the worm exits it, forming a cocoon. Sometimes more cocoons will be formed as the worm uses up all the seminal fluid it has stored. Each one is only a fraction of an inch long and not much wider than the girth of the worm that produced it. (The larger the worm, the larger the cocoon: the giant Australian worm *Megascolides australis* produces cocoons almost as big as a chicken egg.) Regardless of the size, cocoons are shaped like lemons, more or less, and most are brown or dirty yellow. To run across them in the garden soil or the worm bin is a delight: it means that conditions are good, food is abundant, and the time is right for the population to grow.

Once the cocoon is on its own, fertilization can finally take place. If the environment changes drastically—if a heat wave dries out the soil, for instance—the egg and sperm can remain in the cocoon for months before fertilizing. After fertilization, a few more days or weeks might pass before a newborn worm hatches. In some cases, the time could be much longer. (The giant Australian worm cocoons might not hatch for a year.) Usually only one worm hatches from a cocoon, although *Eisenia fetida* has been known to hatch as many as six. Worms don't pass through a larval stage; they emerge from their cocoons as miniature worms, tiny translucent things with what looks like one red vein running through them. I've seen these baby worms in my own bin from time to time. They're hard to spot but it's exciting to find them, these infinitesimally fragile infants who take me back to distant biology classes, and suddenly I can summon up a picture of a capillary or a cilia, something barely thicker than a hair but alive nonetheless, pulsing and moving.

. . .

THERE'S AN OLD SCIENCE classroom poster hanging next to my desk. It's a diagram of earthworm anatomy, probably once part of a set of posters on insects or soil-dwelling creatures. At four feet tall, it's an imposing sight, and it's startling to people who come up to my study and don't expect to see it hanging there. The entire story of the earthworm's life cycle is told in this one captionless poster. At the top, a giant earthworm—as large as a rattlesnake, if this poster were to scale—moves along the surface of the soil, the way it might on a cool, damp night. A couple piles of castings sit nearby, reminding the viewer of its digestive powers. The worm leaves a trail of realistic-looking slime in its wake. Below that, in a cross-section of soil, a worm moves through its burrow. Its head appears at the opening of the burrow; it seems to be interested in a leaf that has dropped to the ground. Next to it, in another cross-section of a burrow, a baby worm emerges from a cocoon, which it has cracked open the way a chick would peck through an eggshell.

In the bottom half of the poster, the background is deep black. Features of the earthworm's anatomy float in this blackness as if they are drifting in space. Here are the reproductive organs, and there are the nerves, the nephridia, the blood vessels, all in shades of red, yellow, pink, and grey. The artist infused the organs with a kind of fifties sensibility, a Space Age hipness. That little bit of impishness in the artist's brush lends an air of excitement, of discovery, to a subject that might seem, at first, mundane. After all, an earthworm is not a complex creature. But when you look at this poster, you realize that a worm is more than a simple tube through which dirt passes, much more. When I lead people up to the attic room where I write, they stand transfixed in front of the poster, as if they are seeing for the first time a map of the stars, or a photograph of the ocean's floor. There is wonder in something so vast, but there is wonder in something so small, too.

ONE NIGHT I SAT on the porch after the fog had rolled in and hidden the stars from view. The garden was all around me, but I could hardly see it. In the dark I breathed the familiar smell of salty ocean air, plus the scent of a jasmine vine that had just started to bloom. Somewhere in that scent of sea and plant life, somewhere at the bottom of it, like a bass note, was the scent of the earth. I went upstairs and opened a window so that scent could drift into the house while I slept.

In the morning I stepped outside to get the newspaper and found the garden bathed in dew. It was early spring and new green shoots were coming up everywhere—on this morning I saw daffodils, irises, and poppies pushing out of the soil. A faint, watery sunlight hit the front porch, just warm enough to make steam rise from the steps. I could hardly bring myself to turn around and go back inside, where the kitchen felt dark and chilly in comparison to the early spring weather outdoors.

There was not much to do in the garden. The summer vegetables were still sprouting on the seed table in the attic, and even the spring flowers for sale at the nursery might not have been sturdy enough to withstand another frost if one came. I invented chores for myself so I could be outside—repairing trellises, turning the compost pile, anything that allowed me to pass a few hours outdoors. Finally I decided to start transplanting perennials. This is the gardener's version of rearranging the furniture—it is a restless and sometimes ill-advised activity. Still, it was only when the plants were mostly dormant and the garden stripped down to its bare bones, that I could see the mistakes I had made last year, the poorly placed flowers and shrubs that should be moved so that the plants would have time to recover before spring began in earnest.

I dug up a half-dozen rose campions that came from my mother's garden and planted them all together near the front of the border. I pulled out a licorice plant and moved it to a bare spot next to the gate. The pincushion flowers I planted last fall turned out to be taller than I expected—they had to move to the back of the border where they could hold their own among the cosmos. As I sunk my shovel into the soil, a few worms turned up. I got down on my knees to look at them more closely. They were mostly endogeic worms, small greyish brown creatures. Occasionally I found a nightcrawler. The blade of my shovel made a clean cut into the soil; I realized that this cross-section of dirt was perforated with tiny holes the diameter of spaghetti strands. There were also slightly larger holes, almost as big around as a pencil. These holes are worm burrows, evidence of the constant movement of earthworms.

The other thing I found as I sifted through the earth was a flat, spoon-shaped worm tail severed from its body. I had cut a nightcrawler in half with my shovel. The sight of that tail all by itself was a little revolting—it reminded me of the dead-finger-in-a-matchbox trick we used to perform at Halloween. I knew the worm would probably grow a new tail and carry on as if nothing happened, but still, it made me squeamish. I didn't like the idea of sinking my shovel into the ground and chopping worms in half, not after I'd grown so familiar with their habits, not after I'd tried to see life from their subterranean perspective.

If I knew that every worm I chopped in half would regenerate into two worms, I wouldn't mind so much. But it's not true that a worm cut in half will turn into two worms. In fact, only one end will regrow (usually the end with the head or the end with the most segments intact), and even then, each species is a little different: nightcrawlers don't regenerate very well, while *Eisenia fetida* will regrow segments quickly and consistently. Generally speaking, you can cut a few segments off the tail of a worm and it'll grow back within a couple weeks. The wound will remain open at the end, with new segments forming a slender tail that expands to its full width only after all the segments are in place. In some cases, if you cut a couple segments off a worm's head, that end will regrow too, one thick segment after another.

This is one of the marvels of an earthworm's life cycle: its ability to grow new body parts, to spontaneously heal from injury. How is it, I wonder, that this creature was endowed with such an ability? I look at the scars left on my own hands after years of meeting up with brambles and sharp pruning shears in the garden. I think about my great-grandmother's hand, which lacked half a digit after an accident with an old cast-iron laundry mangle. Why is it that a worm can regrow most of its body, but we can't replace so much as a finger? I am left with the troubling conclusion that the worm's survival may, in the grand scheme of things, be more important than my own.

EARTHWORM REGENERATION IS one aspect of the worm's life cycle that we do, thankfully, know a great deal about.

An earthworm can be sliced apart in a lab with no regard to the unnaturalness of the setting, and it will respond with the same regenerative vigor that it would in the wild. In fact, so much has been written on the subject that it is clear that these experiments are a favorite of scientists. After all, who could resist the spectacle of dealing a fatal blow to one's subject, only to watch it come back to life? So today we know, for instance, that a worm will never grow more segments than it lost. Whether or not it can fully replace each lost segment depends on how much of its body remains. Many species are remarkably consistent in their regenerative patterns. Take a dozen worms, cut them at the same place, and they will each regrow the same number of segments. It has even been proven that *Eisenia fetida,* if cut between segments eighteen and nineteen, is equally likely to regrow a head or a tail. You could end up with a two-headed or a two-tailed creature that would probably not survive for long. In some cases, a worm cut at this groove has been known to grow what is called "a structure of uncertain constitution": a body part that can only be described as an abnormal head, an abnormal tail, or a "small extra appendage" jutting out from the body at a ninety-degree angle, which may, over time, be reabsorbed into its body.

There are many variations on these regeneration studies; after I read enough of them I could not help but wonder if it was fair to the earthworm to torture it and make it perform in this way. *Eisenia fetida,* for instance, can regenerate after multiple amputations. We know this because researchers have cut the same segments off five or six times and watched them regenerate each time. Some worms have even suffered thirty or forty amputations and regrown segments. And those regrown portions are just as powerful as the original earthworm—if you take only the regenerated portion of a worm's body and cut that off, it, too, can regenerate into a complete worm.

This phenomenon has led some researchers to experiment with transplanting heads or tails from one worm onto another. Like circus animals, the worms oblige and continue to perform. You can cut a tail off and suture it to the head of another worm, and within a couple of weeks, the intestines and nerves will join together and work properly, even if the two ends are rotated at a forty-five-degree angle to one another and then joined. You can take a head from one, a tail from another, and the middle section from a third, suture them all together in the correct sequence, and get one complete worm. Two worms can be stitched together side-by-side, like conjoined twins, and soon they'll grow together and function normally. I've even heard that the first and last segments of a worm can be nicked off and head and tail joined together, so that the worm forms an O. (Admittedly, the worm would not live long in the shape of infinity, with no ability to eat or excrete castings.) The tails from two different worms can be sutured together and the resulting creature will live quite a while, but a worm made from two heads will never be quite right and won't live long.

Accounts of the earthworm's power to regenerate itself go back almost as far as Darwin. Just after the turn of the century, several oligochaetologists published papers explaining their experiments. In particular, they wanted to know if worms could regenerate lost organs, not just lost segments of muscle and intestine. However, in many common species of earthworms, the major organs, including the genitals, are found near the head, where regeneration is spotty at best.

One worm, *Criodrilus lacuum*, keeps its sexual organs near its tail and has shown a powerful ability to regrow those organs when they are cut off. In fact, it will often regrow more testes and ovaries than it had previously: up to twelve pairs, as compared to its usual three. Sometimes it will produce both male and female organs on the same body segment, or it will produce hermaphroditic glands that can perform the male or female function.

Remarkably, on rare occasions just a few segments of a worm have been cut away from its body and a full worm regenerated from those segments, including sexual organs. Scientists have even transplanted the sexual organs from one species into another to try to test whether the worms' regenerative powers make it easier to accept a transplant. Although such occurrences are rare, it was these experiments that led scientists to understand that latent sexual cells exist along the length of an earthworm's body, in every segment, making it possible for a worm to grow sexual organs even after they have been completely cut away.

Even today, little is known about the exact mechanism that regulates the regeneration of lost segments. Understanding regeneration may help scientists explore how human tissue can be regrown. In fact, similar studies on starfish have contributed to advances in human bone regrowth by identifying the particular genes and proteins that enable a starfish to regrow an arm. But the great flurry of earthworm research that followed the publication of Darwin's book had subsided by middle of the twentieth century, when scientists turned their attention to more practical concerns, such as worms' ability to improve soil for farming. Surprisingly little has been added to our knowledge of earthworm biology, reproduction, or regeneration in the last few decades. Even less is known about the very aspect of earthworm life that sparked Darwin's interest in the first place: the earthworm's ability to learn, to make choices—even to think.

FEW PEOPLE SINCE Darwin's day have made serious inquiries into the mental abilities of a worm. Such questions mattered to him, but he was a different kind of scientist, a man with unlimited curiosity and the gift of unrestricted time and resources. Scholars point out that Darwin had a knack for finding something to like about the creature he was studying, some essential quality that he could appreciate, and then his research would be driven by his desire to prove the existence—and importance—of those traits. He could be accused of humanizing earthworms, but what kind of human? In his study of worms, he allowed his pastoral view of honest labor to show through. He admired their humility and their daily toil. His was an egalitarian view, one that defended the contributions of the lowliest worker and recognized the power of the collective. Once Darwin had gone this far, there was only one more logical step to take. He began to look for signs of intelligence among his worms.

Before I read about his worm experiments, I would not have thought to wonder about the creatures' intellectual abilities. When I lift the lid off my worm bin and watch several dozen of my small charges dive away from me, avoiding the light, the thought of teaching them to do anything at all seems ludicrous.

All I ask of them is that they stay in the bin and eat my leftovers. But Darwin asked more of them than this, making inquiries into their senses, their desires, and even their problem-solving abilities.

He wrote to British psychologist George John Romanes, who had published papers and delivered lectures on the subject of animal intelligence, asking if worms could be considered intelligent for the way in which they seized leaves by their apex and pulled them into their burrows. "There may be intelligence without self-consciousness," Romanes replied, and with this assurance, Darwin attributed any number of human qualities to his subjects. In the introduction to *The Formation of Vegetable Mould*, he wrote, "As I was led to keep in my study during many months worms in pots filled with earth, I became interested in them, and wished to learn how far they acted consciously, and how much mental power they displayed."

It is not difficult to imagine the old man passing many happy hours with the colony of worms he established in his study. I know from my own experience that earthworms are companionable creatures—clean, quiet, and hardworking, and for the most part, tolerant of just about any type of experiment, from being teased with a banana skin to having a dozen or so of its rear segments cut off. From time to time I bring a few up to my own study to keep me company. A pot of earthworms on the desk is a pleasant distraction, and Darwin seemed to luxuriate in the uninterrupted hours he spent with his subjects in the name of science.

He began by testing them for sight. He said, with characteristic tact, that they are "destitute of eyes" (another British scientist wrote in 1924 that earthworms have not even "an apology for eyes"), but he went on to illuminate them with a lantern or a candle to gauge their reaction. Although they were often unaffected by candlelight, about one in ten times he observed them retreating into their burrows to avoid light or waving their bodies around like a snake under the spell of a charmer. (I've seen them do it myself; I know what he means. They anchor their tail in the soil and wave their bodies around madly as though they've been seized with the urge to dance or to try walking around on the surface of the earth with the rest of us.) Overall, he observed, they seemed more sensitive to light near their head and less aware of light shining on their tail. (Scientists working after Darwin would discover that the sensitivity to light is triggered by special light-sensitive cells in the skin.)

Darwin continued in this manner, testing them for each of their senses. He blew a loud whistle and found they did not respond to sound. He placed his pots of worms on the piano and played, which sometimes caused them to dash into their burrows. Clearly, he concluded, the worms were sensitive to vibrations. That realization led him outside to beat on the ground, testing the hypothesis that worms come out of their burrows if they believe they are being pursued by a mole. Unable to rouse any worms by pounding on the dirt, he moved on to the subject of smell.

"They were quite indifferent to my breath, as long as I breathed on them gently," he reported. "They exhibited the same indifference to my breath whilst I chewed some tobacco, and while a pellet of cotton-wool with a few drops of mille-fleurs perfume or of acetic acid was kept in my mouth." (Darwin never hesitated to use his mouth as a scientific instrument. Once, as a young man, he was out hunting for beetles when he found himself with a specimen in each hand and another, rarer specimen, within reach. He popped one of the beetles into his mouth for safekeeping so he could grab the third, only to make a new scientific discovery: the beetle he'd placed in his mouth shoots out a noxious liquid when confined.)

"Judging by their eagerness for certain kinds of food, they must enjoy the pleasure of eating," he wrote. He discovered that worms could quickly find their way to a half-rotted cabbage leaf and, after a round of experiments, determined that they preferred green over red cabbage, celery over parsnip, and onion over mint or sage. If they lacked a sense of smell, they still possessed some power to detect the presence of food and move towards it, and then to make careful selections from the food available. Darwin's wife, Emma, watched with amusement. "F[ather] . . . has taken to training earthworms but does not make much progress as they can neither see nor hear," she wrote to their son George. "They are, however, amusing and spend hours in seizing hold of the edge of a cabbage leaf and trying in vain to pull it into their holes."

THE WORMS IN MY own bin don't have to work as hard for their dinner, but over the years, I have noticed that they have some preferences among the foods I offer them. Orange peels and onion skins go untouched for months; I usually deposit them in my outdoor compost pile, along with all the garden trimmings. (Red wigglers could work their way through garden waste, but it would have to be finely chopped for them to make much use of it, so most home composters restrict their worms' diet to kitchen scraps.) A hunk of half-rotten broccoli or cauliflower will also go untouched, probably because it is too large for the worms to work on. They'll wait for it to rot, to break down a little in the bin, before they go after it. They like cooked rice, perhaps because each individual grain is so small that they can make use of it almost right away. Fats, meats, dairy, and spices are inedible to worms and harmful to the bin overall, attracting maggots and other unwelcome creatures. My worms eat a strictly vegan diet.

But fruit—there is the real object of their desire. Once, while I was out of town, I bought an extra carton of worms from a worm farmer I'd been visiting. It would be a couple of days before I could get the worms home, and although the grower assured me that they would do just fine without food for a little while, I worried about them. One morning after breakfast, I chopped up a banana peel and dropped it into the writhing mass of worms packed into their wax-lined cardboard home. The next day, the peel was gone entirely.

Summer fruits are their favorite. I've seen them mass on a mango skin, all competing to get at the sweet fleshy underside. I've even dropped the pit of a mango in their bin, knowing that they'd strip it bare and I'd have to fish it out a month later. Their greatest delight is melon; I've given them the scooped-out insides and the skin, and they've devoured it all, leaving nothing but the seeds, which I've sometimes found sprouting in the castings, as if they'd decided to try a little farming themselves.

. . .

I WISH I COULD say that over the years I've gained some insight into the intelligence of my worms, but the most I've seen them do is act out of instinct or hunger, moving up to higher ground in the bin if water pools in the bottom, or gravitating towards food they like and away from food they don't. If they have an intellect, I don't suppose I've provided much to stimulate it. Still, Darwin's enthusiasm for his worms is contagious. He made a convincing case for their mental and physical capacities, marveling at their industry and at the faint but indisputable intelligence that their habits revealed.

It was an unpopular notion in his time, and perhaps it still is, to draw too many comparisons between humans and earthworms. Do we really want to know that an earthworm can make decisions? Once you realize how many of them live in the soil under your house—ten thousand? a hundred thousand? a million?—do you really want to be convinced that they can think for themselves, that they can take actions? What would happen if all the worms in my bin agreed upon a plan of action—any plan, other than staying at home, eating, and having babies—at the same time? I suppose there's not much I could do to stop them. Like workers going on strike, their power lies in their collectivity.

This comparison of human and animal traits must have been the very thing that troubled Darwin's critics about his work. Just before his death, a cartoon in *Punch's Almanack* showed an earthworm rising from the rubble, growing a monkey's head and tail, standing up on two legs, then carrying a club and loincloth, and eventually morphing into the spitting image of old Darwin himself.

The caption read, "Man is But a Worm."

# European Conquest

The spread of the adventive Lumbricidae into the
temperate zones of the world has, like any good adventure
story, elements of fitness, luck, conquest, failure,
and a good deal of exaggeration.

—WILLIAM FENDER, "Earthworms of the
Western United States," *Megadrilogica,* 1984

SPRING WAS WINDING DOWN and summer was on its
way. The earth warmed a little, and most days, the fog started to
lift around noon. One Saturday I sat outside in the dirt pulling
weeds. It was satisfying work, clearing some space for the sunflowers and zinnias that were only a month old but already tall
and well rooted. After a while I looked up and pushed some hair
under the brim of my hat. It was a fine feeling to be outdoors in
the summer, with the fog slowly burning off and the sturdy
green plants reaching for the sun. As I sat and surveyed the bed
I'd just weeded, a fat red worm moved across the soil towards
me. My heart leapt a little at the sight of it. I dropped my spade
and picked the worm up, rolling it over in my hand. It was a
large specimen, red on top with a translucent underbelly, a barely
developed clitellum, and a flat, spoon-shaped posterior. *Lumbricus terrestris,* the nightcrawler.

I have an old British earthworm textbook that includes a section on "Mesmerizing Worms." The author contends that crabs, lobsters, fish, and snakes all like to be stroked; it is, he claims, a well-known fact. He described stroking a nightcrawler on its back one day and finding that it "instantly settled down to be petted." A few days later, he picked up another one, and although it squirmed around at first, it later "became perfectly quiet when stroked, and lay still afterwards as if it were waiting for further play."

Until I read that passage, it had never occurred to me to stroke a worm. I would not have thought that a worm longed for physical contact, although I'd seen them mass together in the bin in a way that showed they were not afraid of crowds. But why would a nightcrawler tolerate exposure to air and an unfamiliar human touch? And how would I know if the worm was mesmerized or merely playing dead?

I pulled my glove off and ran a finger along the worm's back, from head to tail. It stretched as I touched it, lengthening each segment under the pressure of my finger. Still, it didn't seem to be trying to escape, just extending itself. I stroked it a few more times, slowly, guessing that a faster stroke would only alarm it. Soon it was perfectly still in my hand.

What can I make of this interaction between me and the nightcrawler? Surely it doesn't know a thing about me. If it could see me, it would have every reason to be frightened. I can only guess that everything above the surface of the earth is outer space to a worm, and that humans are aliens. Why would it tolerate me stroking it? Maybe the pressure on its body reminds it of the comforting closeness of an earthworm burrow, where the sensation of gritty earth sliding along the skin must feel like home.

An earthworm's skin is fragile; I didn't see the point in handling it any longer. But when I looked around the garden, I couldn't tell where it came from. There was no burrow nearby that I could see. I set it down in some loose soil and sat next to it, waiting for it to bury itself. At first it didn't move. Maybe it was still mesmerized. Then it pushed a little soil aside and found a way back into the earth, its flat iridescent tail waving as it ducked away from the light and continued its work.

As FAMILIAR AS THAT nightcrawler may seem, it is not indigenous to North America. In fact, many of the worms commonly found in American garden soil today are not native. Canada, the northernmost parts of the United States, and northern European countries such as Sweden and Norway were covered by ice and snow during the last Ice Age about ten thousand years ago. If earthworms existed there before the advance of the ice, there were none to be found after. And the native worms that lived below the southern limit of the glaciers have gradually been displaced through the disruption of their habitat and the introduction of exotic (non-native) species. Those native worms are still out there, but they are poorly understood and increasingly difficult to find. Dig up an earthworm from any backyard in the United States, and that worm will mostly likely be non-native.

Lumbricid earthworms like the nightcrawler followed the route of immigrants into North America. Some ecologists have called these worms "weed species"— creatures, like cockroaches and flies, that follow humans wherever they settle. They have also changed the nature of the landscape, and the possibilities for agriculture, everywhere they've gone.

European immigrant farmers coming to the United States might not have thought to deliberately transport such a ubiquitous creature as an earthworm to the new world, but transport them they did—in the soil of potted plants, the ballast of ships, the hooves of horses, and the wheels of wagons. Whether they came as stowaways or invited guests, there is no doubt that foreign species of earthworms, namely *Lumbricus terrestris,* the nightcrawler; *Eisenia fetida,* the red wiggler; and *Aporrectodea caliginosa,* the field worm, came west with settlers and helped turn the already rich valleys and plains of the Midwest into some of the best farmland in the world. In two hundred years, European settlers managed to transport earthworms nearly three thousand miles across the continent, a feat that would have taken the worms about 1.5 million years to accomplish by themselves had they simply been released on the East Coast and allowed to migrate towards California at their leisurely rate of a few yards per year.

When these new species arrived in North America, they found a welcoming climate: rich, damp soils; a comfortable pH level; and temperate weather. There is almost no part of the United States where European earthworms have not made a home for themselves. Over the last few decades, earthworm censuses have found over a million worms per acre in Geneva, New York; Frederick, Maryland; and LaCrosse, Wisconsin.

I can only assume that the worms in my backyard are recent immigrants themselves. Eureka was established late because of its inaccessibility. Dense redwood forests made the area hard to reach by land and it was barely visible from sea, where only a narrow opening led into the bay. As timber and fishing industries began to flourish in the mid-1800s, settlers began cutting down forests to make room for a town.

My house was built around the turn of the century by a family—the Chapmans—who had owned the land since about 1880. When they dug a well in my backyard to serve the three houses they would build on this block, they might not have found any nightcrawlers; the European worms were also recent arrivals and probably hadn't established themselves yet. For that matter, they might not have found any worms at all, considering that they were building on cleared forest land. Not much is known about the native species that lived in the redwood forests at the time. Whatever species did exist probably disappeared as the forests were cleared to make way for a grid of streets and houses. The nightcrawlers in my garden could have arrived a hundred years ago in the potting soil of Mrs. Chapman's rhododendron, or they could have been introduced over the last few decades when the previous owners of this house returned from a fishing trip with extra bait.

ALTHOUGH A GOOD DEAL of forest was cleared to make way for the town of Eureka, redwood trees can still be found in people's backyards, and the town is surrounded by wilderness. In fact, the northern entrance to the Headwaters Forest is only about a fifteen-minute drive from my house. I go there from time to time in the summer. There's a comfortable, paved trail leading into the woods, and it makes for a nice walk in fair weather. Sometimes I'll see hawks and songbirds there; when it's damp I may see a red-legged frog or a dusky yellow banana slug. But now, when I walk through the woods, I wonder about what I can't see. I wonder about the worms in the forest.

It seems as though no one around here is neutral when it comes to the Headwaters. To timber companies the forest represents jobs and lumber, and to environmentalists it represents a priceless resource: some of the last stands of pristine old-growth redwood forest. Few have forgotten the protests that took place a few years ago to save this forest. After all, this is the place where Julia Butterfly Hill lived for two years in a tree named Luna, and where police officers swabbed the eyes of antilogging protestors with Q-tips dipped in pepper spray, prompting a civil rights lawsuit that is still working its way through the courts. At the end of all the protests and negotiations, the federal Bureau of Land Management and the State of California joined together to purchase the 7,400-acre forest, allowing them to halt logging and protect such endangered and threatened species as the marbled murrelet, spotted owl, and coho salmon.

Most of the Headwaters forest is off-limits to visitors. From the southern entrance, forest rangers lead small groups on a four-mile hike along what used to be a logging road. Although there are some old-growth trees along the hike, most of the forest near the trail has been logged at one time. At the northern entrance— the one closest to my house—people can walk about five miles into the forest, but along the way they'll see more maples, willows, and alders than they will redwoods. The most pristine parts of the forest begin where those trails end. This is where the marbled murrelet makes its home. Most tourists who visit will not see the endangered shorebird, or the old groves it inhabits, where ancient trees reach three hundred feet into the sky.

I've never been into those restricted areas, but I've hiked through plenty of redwood forests and I can imagine what these are like. Like tropical rainforests, the trees in redwood forests tend to regulate the climate. The tree canopy shelters the forest from sunlight and filters rainwater as it comes down, keeping the forest cool and moist but never drenched. I have walked into a redwood forest during a rainstorm, only to find I didn't need my umbrella once I was under the canopy. Even on the hottest afternoons, under the trees it is always chilly and damp. Some trees are so enormous that five people can't link arms and reach around them, and they are so tall that it is impossible to stand at the base and see the top.

A redwood measures its life span in centuries, not decades. Time seems to stand still inside a redwood forest—it is easy to imagine that you have dropped into some distant past, centuries ago, when these forests ran in a narrow band along much of the West Coast—but at the same time, life begins again so quickly that you can almost watch it happening. When lightning strikes a tree, a fire burns down through the center of it. Around it, dozens of new redwoods spring up from the cinders. When one falls, it becomes host to mosses, ferns, and young redwood saplings. Even the smallest creatures seem primeval, otherworldly: I have seen red-legged frogs looking up at me from a damp creek bed, and banana slugs, mustard yellow and longer than my finger, move across the trails at an infinitesimally slow pace. A redwood forest is powerfully alive. When I visit one, I put my hands on the largest tree I can find, and I almost believe I can feel a heartbeat. Even in my most pragmatic moments, when I acknowledge that I live in a house built of redwood, filled with wood furniture and thousands of books printed on wood-fiber paper, I can't help but shudder at the thought of even one of these trees being cut down and hauled out of the forest.

That's why it always gives me an uncomfortable feeling to walk into the Headwaters from the northern entrance and see the remains of the old logging operation. The town of Falk once stood on a hillside about a mile into this forest. It was a lumber town, owned by the Elk River Lumber Company and occupied by about four hundred people at its peak. People lived in Falk from about 1884 until just after the Great Depression. By 1940, it was a ghost town, complete with an abandoned general store that held nothing but empty shelves and old receipt books. Hikers used to go out there looking for artifacts, and the empty buildings attracted plenty of campers over the years. In 1979 the last of the old buildings was knocked down. The forest has begun to move back in, reclaiming its territory, covering up the signs of human occupation. A pile of brick and rubble—perhaps a house or an outbuilding?—has sunk into the forest floor. Moss and fern almost completely obscure the pile of bricks from view. Redwood saplings have started to spring up among them. Over time the remnants of this town will be completely buried under the forest floor. Right now, little remains but the plants that Falk residents brought in to plant around their homes. Garden plants like yew trees, daffodils, and breath of heaven shrubs can all be found along the trails.

I got to thinking about those plants after I'd been to the Headwaters a few times. If earthworms migrated across the country in the roots of plants, as many believe, does that mean that the Falk townspeople brought earthworms into the Headwaters forest? Is it possible that the exotic plants growing alongside the trail are just the tip of the iceberg? What would it mean to this forest—or any forest—if an unfamiliar species of earthworm moved in?

I drove out to the northern entrance one Sunday afternoon in May and walked into the forest, taking a self-guided tour brochure with me. The second stop along the tour was a pair of yew trees that once marked the entrance to the home of Loleta and Charlie Webb, Falk's caretakers. The column-shaped trees were strikingly out of place in the forest. Between them, a path led to the site of the caretakers' former home, but the path was off-limits to visitors. A short post held a plaque that explained the site's significance. At the base of the post were a couple of bricks. I kicked over one of the bricks. Under it, three earthworms squirmed out of sight.

I grabbed a stick and dug them out of their burrows. They looked like ordinary pink worms to me, smaller than a nightcrawler, lighter in color than the red wigglers that lived in my worm bin. I couldn't begin to guess whether this was a native or an exotic worm. Sam James would know. I wrapped my hand around the worm and kept walking. A little while later I came to the spot where the old mill had stood. The cookhouse and loggers' homes once stretched up the hillside behind the mill. No one was allowed to climb up the hillside, but I turned over a few logs near the trail and found a couple more worms. The next morning, I shipped them off to Sam James and waited for the results.

I knew that the worms I sent to Sam were in no way representative of all the worms living in the forest. One scientist wrote an article about earthworms of the Pacific Northwest several years ago in which he warned that "many species preferentially inhabit game trails and other compacted soils. . . . Do not fall into the habit of collecting only the easiest habitats." He suggested getting at least twenty yards away from roads before collecting specimens, but I didn't dare wander that far off the trail in the protected Headwaters forest. Even if I had, I might not have found any native worms. Another taxonomist wrote that the absence of native worms in some areas of California "now shows that they were unable to compete there once the adventive Lumbricidae arrived." If native worms were to be found, they were most likely deep in the pristine old groves, far away from human interference from the likes of me and the long-gone Falk townspeople.

Sure enough, Sam reported a week later that the worms I'd sent belonged to the Lumbricidae family. I'd uncovered *Lumbricus rubellus*, redworm. The other worm I'd dug up was *Aporrectodea caliginosa*, the common field worm that also lives in my garden. Both are non-native worms, and both thrive in the damp soil of the Pacific Northwest. Because they're both so common to this part of the country, it is impossible to tell how they arrived in the forest. They could have moved in along with the residents of Falk, or they could have come in later, as heavy equipment rolled in to pave the trail or raze the last of Falk's buildings. Even a muddy hiking boot could have deposited a worm cocoon along the side of the trail.

What does it mean that these non-native worms inhabit the forest floor? It's a question that ecologists are just beginning to explore. An invasive plant species can move into a forest and displace native plants, change the food source and habitat available to birds and other denizens of the woods, and even—through the gradual thinning of native trees and shrubs—alter the climate inside the forest. One could speculate that exotic earthworms displace native earthworms by consuming their food source or pushing them out of their habitat. A native earthworm may have a symbiotic relationship with a particular native plant, or it may be a food source to a specific bird or snake. It may help maintain a particular balance of nutrients in the soil, one that forest plants rely upon. If that earthworm disappears, is it possible that the forest itself could change?

Answers to these questions are not easy to find. Sam James wrote that there are "serious deficiencies" in our understanding of native North American species and lamented the fact that there may not be "sufficient social or political interest to foster conservation of earthworm biodiversity." After he identified the worms for me, I called him to ask if it was possible for a non-native species to harm the Headwaters forest.

"Hard to say," he said. "But it's an interesting question. You don't really think of earthworms as being destructive, but that's what brought me down to the Philippines the first time." He'd heard that earthworms were destroying rice crops there and offered his services to identify the worms and help local farmers come up with a solution. He first traveled to the Banaue rice terraces in the Ifugao province of the Philippines in 1999.

The terraces, which have been in place for upwards of two thousand years, climb up the sides of green mountains that rise about four thousand to six thousand feet above sea level. The terraces are flooded during the growing season, but when the season ends, the fields are drained and the farmers grow other crops along the terraces. "It's a modern economy there," Sam said. "They can't rely on rice alone to support themselves, so they drain the terraces and plant vegetables and flowers." Sam and his crew arrived to learn that foreign earthworm species—some from South American and some from Asia—had come into the area as roads were built and new plants were imported. They were active burrowers that filled the terraces with holes as soon as they were drained. "It's like Swiss cheese out there," Sam told me. "The floods don't hold. Used to be they'd keep them flooded year-round, and the worms never had a chance to get in."

That's not the only worm problem facing these farmers. "There's another worm that lives in the terrace walls itself," Sam said. "Some of the farmers live out there on the terraces and these worms are attracted to the garbage that piles up. The worms are building burrows that break down the terrace walls. It's a native worm that's hard to identify because it has no sexual organs. We rely on those organs and their location to identify most species of worms. But this worm is parthenogenic. It clones itself. Hard to say for sure what it is."

There's no easy solution to this problem of earthworms in the rice terraces. It points out, once again, how small acts have unintended consequences. "There are plenty of scientists out there studying the ecology of invasions," Sam told me. "Until recently, earthworms weren't really on their radar screen. But you never know what the consequences will be of bringing a few worms into a new environment."

I mentioned the eucalyptus trees that grow along the California coastline. They're not native, they grow like weeds, they kill anything that lives underneath them except other eucalyptus, and they are a serious fire hazard because of their high oil content. They were brought here from Australia over a hundred years ago as a timber source and windbreak. "I think most people can relate to the idea of an invasive exotic. Everybody in California knows about the problems with eucalyptus trees. But they're big and obvious," I told him.

"Exactly," he said. "But what about the earthworms that came over in the roots of those trees? Nobody thought about that."

I asked if he had ideas for ridding the rice terraces of the worms that had colonized there. After all, you can't just round up worms and send them home. "That's the hard part," he said. "There's not much you can do. You can't poison them without poisoning the soil. The fact is, the worms may be filling the terraces with holes, but it was an economic decision to drain them in the first place. It may take a government program to pay the farmers to grow rice year-round the way they used to, just to preserve that way of life.

"But you know one thing that will get those worms out of the ground? Wasabi. You know, the really spicy green stuff you eat with sushi. They can't stand it; they'll do anything to get away from it. But can you imagine what that would smell like? A whole field of wasabi?"

He starts laughing, the ironic laugh of someone who has seen earthworms get the better of two thousand years of rice farming in one country while transforming the soil and making productive farming possible in others. Sam James is perhaps one of the few people who gets the joke: earthworms can change the course of human civilization and do it all silently, in the dark, unseen.

ONE WORM THAT I didn't find in the Headwaters forest was the common nightcrawler, *Lumbricus terrestris*. While they thrive in my garden and are often found in rich soil wherever humans live, they have a low tolerance for the acidic soils of coniferous forests like the Headwaters. They do seem to be fond of deciduous forests, like the hardwood forests along the East Coast and near the Great Lakes, where annual leaf fall makes for a thick pile of worm-friendly compost. A few days after Sam James called me with the news about my Headwaters earthworms, I heard about a couple of scientists at the University of Minnesota who were studying the earthworms' destruction of hardwood forests. Before long, I was on my way to Minneapolis.

# In the Forest

In woods again, if the loose leaves in autumn are removed,
the whole surface will be found strewn with castings.

—CHARLES DARWIN,
*The Formation of Vegetable Mould*, 1881

I ARRIVED IN MINNESOTA on a clear, chilly day. It was
May, but there was still snow on the ground. I met researchers
Lee Frelich and Cindy Hale at the university, and they agreed
to show me the forest where earthworms were causing such a
problem. It took less than an hour to get there from campus, and
on the way Cindy told me that most people react with genuine
surprise when they find out what she's working on.

"People always tell me, 'I thought earthworms were good for
the soil,' and I tell them that they are, in some settings. But these
European worms have invaded a forest that evolved without
them over the last ten thousand years. Remember, this whole
area was covered by glaciers until the end of the Ice Age. When
the ice melted, there wouldn't have been any worms."

Cindy didn't set out to study worms. She began her work as
a forest ecologist, and in the process she and her colleagues noticed
something changing in Minnesota's forests. The understory was

dying. Ferns were disappearing, wildflowers had all but vanished, and young tree seedlings couldn't take root. Forestry experts couldn't figure out what was happening, but they knew the forest couldn't survive without this critical understory of small plants.

Lee, her doctoral supervisor, recalled how puzzling the situation was at first. "A couple plants really took over in some of the forests and replaced all the ferns and wildflowers that used to grow there. Pennsylvania sedge, which looks like grass, and one other plant, jack-in-the-pulpit, started to carpet the forest floor. People used to call me all the time asking what was going on in our forests. Nobody knew."

"Then somebody published an article about the changes that take place in the ecology between urban and rural areas," Cindy said. "It mentioned, in a kind of offhand way, that increases in earthworm populations might be causing changes in understory plant populations of New York forests. That's when it finally occurred to us to go out into the forest with a shovel and dig."

What they found was nine species of worms, including *Lumbricus terrestris* and other exotic species. Darwin noted how clever the nightcrawler was when it came to pulling leaves and pine needles into its burrow. But even he failed to realize how this efficiency, this special skill, could be destructive on such a large scale.

Earthworms, the Minnesota research team has learned, can — and do — consume the entire leaf fall of a forest in a single season. Small plants and tree seedlings flourish in the damp, sweet-smelling, slowly decaying layer of forest floor. This layer, the duff, is built up over many years. It contains leaves and other organic matter in all stages of decay. Many of the native plants that once flourished in the forest produce seeds that have intricate germination strategies. A seed might take two or three years to germinate, going through a complicated cycle that depends on this spongy duff layer. Now that the forest floor is bare, most small plants have simply disappeared.

"We've seen a loss of eighty to ninety percent of all understory plants in some areas," Cindy said. "That's where we find the most earthworms. They just expand their population to fit the available food source. They multiply until there are enough of them to eat all the leaf litter on the soil's surface. And the ten or twenty percent of plants that do survive? The deer get those."

I could hardly believe what she was telling me. I thought about the redwood forests back in California. I couldn't imagine those forests without ferns, native columbine, moss. "It's just so counterintuitive to think of earthworms as pests," I said. "Does it ever sound strange to hear those words coming out of your mouth?"

"Not anymore." She'd been studying this problem for four years. "But yeah, that's what I have to remind everybody. Earthworms are great in a compost pile. They're wonderful for agriculture. They do till the soil. They do add nutrients. They do all the wonderful things everyone has always believed them to do. But when they move into a forest that has evolved without earthworms, they can actually have negative effects on the native plants."

I told her what I knew about the earthworm migration from Europe: that they arrived in potted plants, ship ballast, even in cocoons on the soles of shoes. How did they get into Minnesota's forests? Have they just slithered in that direction over the years?

"Think about it," Cindy said. "You've been at the lake fishing all day. You still have a few bait worms left when you're finished. What do you do with them?"

"Dump them in the soil," I said slowly, feeling a little uneasy as I thought about the bait stand worms I've dumped in my own garden.

"Exactly. That's how they're getting in. That, and ATV tires that have mud caked in them, and fill dirt people bring in when they build a cabin, and potted plants for landscaping projects. People are bringing them in. They couldn't move this fast on their own."

Lee added, "Wilderness managers have known for a long time that you shouldn't just bring a new species into an environment like this. There's a threshold effect that can happen in an ecosystem. It can tolerate just so much change, then it snaps like a rubber band. There are wilderness areas in Michigan where they've banned live bait since 1965. They might not have known exactly what the worms could do to the forest, but they knew better than to bring exotic species in where they could reproduce unchecked."

Cindy and Lee have encountered their share of skeptics. "After all," Cindy said, "We're working against generations of— well, of common knowledge! It just makes sense that worms are good for the soil. But when I take people out to the forest, and they see what's happening, I can make believers out of them." Lee told me that when he first applied for funding to do this research, he met a lot of resistance. "Earthworm researchers are working so hard to prove all the benefits that worms bring to the soil. Nobody wanted to fund research that showed they might be a pest.

"A couple years ago I went to an ecological conference on hardwood forests," Cindy added. "I sat in the back of every presentation and at the end I raised my hand and asked, 'Do you have worms in your forest?' I must have sounded nuts. But you know what? Nobody knew. I got one call from a guy who worked in the Department of Resources in New Hampshire. He'd been telling everybody for years that they should try to keep worms from being imported into the forest. Nobody believed him. There just aren't a lot of people looking into this yet."

Then she said something that no earthworm scientist had said to me. "They can be so beneficial, or so destructive," she said. "They are literally ecosystem engineers. They are at the very base of the ecosystem. Their actions drive everything else that happens. And yet there are a lot of ecologists out there who pay no attention to earthworms at all."

We arrived at the Wood-Rill forest, a 150-acre preserve outside the Twin Cities. Buds were just beginning to appear on the maple trees. Over the weekend an unexpected snowstorm had left several inches of snow in the forest. I wasn't surprised that the forest floor was still brown and bare, but Cindy and Lee told me that a carpet of young green plants should have sprouted underfoot by now.

"You're not used to seeing this forest," Cindy said, "but let me tell you: this is shocking to us. It looks like the forest's been nuked. There's no trillium or wild ginger, no false Solomon's seal, no bellwort. And look at this." She kicked aside a layer of brown leaves on the forest floor. "These leaves fell last autumn. Usually there would be layers and layers of leaves from earlier years underneath. The forest floor used to be so spongy, it felt like you were walking on a mattress. Now look what's happened."

I knelt down and peered at the black earth under the leaves. It looked exactly like the castings I collected from my earthworm composter. "This is nothing but worm castings!" I said, and she nodded. I scraped the ground with a stick and turned up a tiny pink worm.

"Probably *Aporrectodea rosea*," she said. "It's a little endogeic worm. That's one thing we're just starting to figure out—what role each species plays in this invasion. The epigeic worms seem to come in first and eat all the rotten leaves. Then we start to see soil-dwelling endogeic worms, like this one, then the anecic worms like the nightcrawler—the ones that pull fresh litter into their burrows—come along behind them and take care of the stuff that hasn't even started to rot. All these leaves will be gone by summer."

I told Cindy and Lee that I couldn't believe how fast the worms have moved in, or how many different exotic species she'd found.

"You're not the only one," Lee said. "When we applied for funding a couple years ago, we were nearly denied in the first round because the reviewers—all scientists—simply didn't believe what we were saying. They just couldn't believe that all these earthworms were exotic, or that they were capable of doing this kind of damage."

"But look over there," Cindy said. We were standing on a

slight rise at the edge of the forest. Off in the distance was the brilliant green of a golf course. "They probably brought in a hundred acres of sod to build that course," she said. "Imagine all the worms in that sod. And I buy worms at the bait stand for my research. One time I found four different species of worms in one container of fish bait. They're coming in from everywhere."

Lee and Cindy are only just beginning to document the extent of the problem. They've got graduate students doing field experiments that are not too different from the experiments Darwin did. One student sorts the fallen leaves, placing sugar maple, basswood, and oak leaves in separate piles on the ground, secured by wire cages. He visits the site regularly to monitor the food preferences of the worms. At a "leading edge" site near Duluth, a place where worms have not fully invaded the forest, plant life is carefully monitored so that data can be gathered on changes in growth and germination as worm populations advance. And a series of experiments tracks the food preferences of individual worm species living in pots of soil with particular leaves added as food.

After we'd been walking awhile, we came across a high wire fence in the forest. "This is one edge of the deer excluder," Lee said. "We realized that some plants do survive the worm invasion—maybe twenty percent. But those plants are eaten by deer. Oddly, deer may be the key to all of this. If we can keep the deer out, the plants might just have a chance."

The forest didn't look any different inside the deer excluder. "Is it working?" I asked.

"Look at the young trees out here." Among the bare trunks of mature trees I started to pick out a few saplings only five or six feet tall.

"Those are about twenty-five years old," he said. "There's nothing younger than that around here. The deer have gotten all of them. But just inside the fence, we've got one- and two-year-old trees."

Sure enough, when I looked closer I realized that young trees, each little more than a few twigs with fragile leaf buds at the end, were sprouting. The forest still didn't look lush and green, but to an ecologist, this was real progress.

On the way back to the car, I asked Cindy about the animal population of the forest. She had described the forest floor as a crucial component that affects everything else. Do other animals depend on the forest floor for their survival, animals that the earthworms have displaced?

"Sure," she said. "As worms come into the forest, we see a shift from voles and shrews to mice. There are all kinds of frogs and other amphibians that live in that duff layer. And there's even a ground warbler that nests in the forest floor. It's called an oven-bird because its nest looks like a little oven. We're only just starting to look at that."

It's hard to believe that a creature as small as an earthworm could push ground-dwelling birds and animals out of a forest, but this is exactly what they think is happening. That's not all: insects that live in the duff layer—including microscopic creatures such as springtails—may be disappearing before they have even been identified and described. The change in soil texture could lead to erosion, especially in the summer when water runs across hard, bare ground in sheets. Even the composition of the soil can change; the presence of earthworms can lead to an increase in bacteria and a decrease in fungi populations, which could in turn affect which plant types proliferate and which struggle or fail entirely.

I looked up at the bare branches high above me. How could an earthworm push something as enormous as a tree out of the forest? But there was no doubt that they were accomplishing something out there in the woods, in their slow, methodical way. I thought about the logging protests in the redwood forests back home, and the tree sitters living high in the canopy. The fight to save those forests happens aboveground. It is a battle for the part of the forest that we can see: branches, leaves, tree trunks. But the fate of this forest in Minnesota lies entirely with the part of the forest we can't see: the dark underground.

I ASKED LEE AND CINDY what they thought should happen next. You can't put up a fence to keep out earthworms. What do they want the public to do?

"We've got to educate wilderness managers and the fishing community," Lee said. "They've either got to ban live bait altogether or at least stop dumping their leftover worms on the ground at the end of the day. We need to manage the deer population, since they're grazing on the few plants that do survive. There are some forests in southern Minnesota that are completely invaded by earthworms, but they've been able to keep the deer count low, and the understory's managed to survive."

"What about farmers?" I asked. "You've got farmland adjacent to forests. The farmer wants to cut back on chemicals and build up the population of earthworms in the soil. What should they do?"

"Frankly," Lee told me, "the forested areas near farms are already invaded by worms. There's no reason for farmers to cut back on their use of them. It won't make any difference now anyway. But even in those areas, there are things we can do. We can replant the understory, but we've got to learn a lot about our native plants first. Cindy tried to grow some seeds for her research and found out that they have to go through two or three winters before they germinate."

"You haven't got that kind of time," I said.

"Not if you're a graduate student trying to finish your dissertation," Lee said. "She decided to grow columbine instead. They'll bloom in one season."

"Look," said Cindy, "worms by themselves only travel a few meters a year. You do the math—it would take them about a hundred years to travel a quarter-mile. If we can keep people from bringing them into the forests, we've got some time. There's plenty of forest land that is still worm free. We can keep it that way for quite a while."

But Cindy knows what she is up against. "The U.S. imports millions of dollars' worth of earthworms every year from Canada. That's a lot of worms. And I'm not opposed to earthworms generally. I've got two worm composters at home myself, but I try not to add any worms to my garden soil. I tell people to take their castings, put them in one of those freezer bags, and put them in the freezer for at least a week before they add it to their garden soil. It won't hurt the soil microbes, but it will kill all the worms."

I couldn't help but picture the look on my husband Scott's face if he saw a bag of earthworm manure in the freezer. "Do you have earthworm castings in your freezer right now?" I asked.

"I do," she said. "It works great. You should try it."

I told her I'd think it over. The fact is, I know there are European worms in my soil already, and at this point, I'm not worried about adding more.

BACK HOME, I LEANED against the window in my study and thought about what Lee had said. There's no reason for farmers to feel guilty about the non-native earthworms living in their fields—those vigorous European species have traveled, alongside humans, to just about every square mile of land in this country and in many places around the world. This shift in the earthworm population has already happened; it would be futile for a farmer or a gardener like me to be concerned about a few more European worms making their way into the soil. Still, as I looked out over the grid of streets that form downtown Eureka, and Humboldt Bay beyond that, and finally looked up to the redwood forests that rise along the hills around the bay, I couldn't help but wonder where, in the fragile balance of human needs and ecological concerns, the earthworm might take us next in its unintentional tipping of the scales.

# Stalking the Giant Worm

A worm is as good a traveler as a grasshopper or a cricket,
and a much wiser settler. With all their activity these do not
hop away from drought nor forward to summer. We do not
avoid evil by fleeing before it, but by rising above or
diving below its plane; as the worm escapes drought
and frost by boring a few inches deeper.

—HENRY DAVID THOREAU,
*A Week on the Concord and Merrimack Rivers*, 1849

WHEN I BROUGHT my first batch of earthworms home
and introduced them to the worm bin, I could hardly wait to
get to know them, to become familiar with their habits, their
likes and dislikes. The thought of hundreds of earthworms in a
bin on my porch proved to be an irresistible attraction. I knew
I should leave them alone for a few days so they could get settled—the instructions that came with the bin were quite ex-
plicit on this point. Earthworms, through their digestion, help
create the kind of microbial community that they prefer. The organisms in their castings flourish and multiply until the balance
of bacteria, protozoa, and fungi is just right. A responsible earthworm farmer will leave a new group of worms in peace for weeks
while they adapt to their new home.

My bin came with a brick of compressed shredded coconut fiber, which expands once it has sat in a bucket of water for an hour or so. It loosens into a damp shaggy mass, losing its brick shape and taking on the consistency of peat moss. This is considered the ideal bedding for a new small-scale worm bin; it gives the worms something dark and damp to bury themselves in. They will eventually eat it. After that they will live in their own castings and won't require any special bedding.

I should have left them undisturbed in this bedding for a week or more, but I just couldn't, not even for an hour or a day. I was too curious about what they were doing in the dark, damp confines of their new home. I wasn't supposed to feed them right away—they needed a few days to get settled in before they were ready to eat—but I dropped a banana skin in the bin anyway. Several times a day I would steal out to the porch, lift the lid, and churn through the bedding with a garden fork. The worms were limp, puny things that lacked the wherewithal to duck away from the light. They needed to fatten up, start laying eggs, explore the boundaries of their new world. I'd have to stop torturing them with the garden fork if I ever wanted to establish a successful colony. Looking back on those early days and the unrelenting, even cruel, nature of my own curiosity, I realize that perhaps it is best that we do not dig too enthusiastically in search of undiscovered species of worms. Our investigation of them could, in the end, be their undoing.

I should probably leave the worms in my own garden undisturbed, but I can't resist. I own a pocket loupe, a magnifying glass about the size of a quarter that retracts into a metal case. Garden supply companies sell pocket loupes to allow gardeners like me to peer closely at spots on leaves, tiny insects, and near-invisible seeds that are too difficult to handle and space evenly without some magnification. I carry mine with me when I'm in the garden and I use it to look at the worms I turn up in the soil. But even a tenfold magnification isn't enough for me to see all the fine details on a worm's body, and after all the time I've spent studying its anatomy, it has not gotten much easier for me to tell different species of worms apart.

Anytime I find a worm in my garden, I turn it over, looking for pores and bristles. I try to count the number of segments between its head and its clitellum (which is another way of distinguishing between species), but the worm stretches and twists and arches away from me, and I lose count. I hold it in one hand and stroke it with the other, trying to mesmerize it, but that leaves me without a hand to hold the loupe. Eventually I give up and drop the worm back into the soil, uncertain of what I've seen. Now that I'm in the habit of picking up worms and studying them up close, I have started to have a recurring dream of magnified worms that appear to be almost a foot long and as big around as my thumb, with rubbery red hairs protruding along every segment. They look like the sort of creatures you might pull up in a fishing net and toss quickly back into the ocean before you even have time to wonder what it was. Even in the dream I turn away from them, as if I've seen something I shouldn't have.

GIANT WORMS AREN'T just a creation of my dreams. There are a few species of giant worms around the world, but I'll probably never see one. I'll certainly never encounter one in my backyard, where the building of homes and roads, the digging of a well, and the constant traffic of humans long ago would have driven any away. And I'm not likely to see one in a zoo. A giant worm wouldn't survive long in captivity, and although a few have been rounded up and put in glass cages, they are nearing extinction now and will no longer be taken from the wilderness for our entertainment.

There are some giant worms up the road from here, in Oregon and Washington State. Native to prairie grassland, they are increasingly rare due to the spread of cities. Even agriculture pushes out fragile native worms that are best adapted to undisturbed soil. Perhaps one of the most unusual worms at risk of extinction is the giant, flower-scented worm that lives in the Palouse region of southeastern Washington State, an area better known for its gentle rolling hills, spectacular rivers and waterfalls, and rich agricultural land. (Locals boast that it is one of the richest wheat-growing regions anywhere, not to mention the lentil-growing capital of the world.)

Few people know that Palouse is also home to the giant Palouse earthworm, *Driloleirus americanus*. Or perhaps I should say that it used to be home to the giant worm. Not one sighting has been reported in over twenty years, and earthworm researchers are pretty confident that if one was found, it would be reported to the local university and might even make the evening news. These worms are pinkish white, reach two feet or longer when fully extended, and are covered with tiny setae that they use for locomotion. They excrete a milky white fluid, the coelomic mucus, that smells like flowers—lilies, to be exact.

The giant Palouse earthworm has been spotted only a few times in the last century. A researcher named Frank Smith found one in 1897, decided it must be a relative of *Megascolides australis,* the legendary Australian worm, and named it *Megascolides americanus* in its honor. It was renamed *Driloleirus americanus* once it became clear that the Palouse worm and the Australian worm had little in common apart from their size— and even there, the Australian type was quite a bit longer than the Palouse worm.

A few more Palouse worms were seen in the thirties and again in the seventies and eighties. Recently, expeditions to un- earth the giant worm have turned up nothing but a few Eu- ropean species. Washington State University even recruited a group of schoolchildren to head to a deserted spot outside of town where it might be found. A day of digging turned up plenty of European worms, but no one caught a glimpse of the giant Palouse worm.

These giant worms attract a small but enthusiastic group of supporters. Another Pacific Northwest worm, the giant Ore- gon worm *Driloleirus macelfreshi,* has disappeared as quickly as its Palouse cousin, but not before a family of earthworm scien- tists managed to name the species and even capture and preserve a few specimens.

Dorothy McKey-Fender, an oligochaetologist, and her husband discovered and identified the giant Oregon worm. Now eighty- five years old, she recalled the days that she and her late hus- band, Kenneth, spent outdoors looking for the legendary worm.

"We'd just take the kids with us," she told me. "That's what we did on the weekends. I was always interested in bugs myself as a child. I knew my kids would enjoy it, too, and my son, William, really stuck with it. He still helps me with my work."

That work includes curating a collection of preserved worms collected over a lifetime. McKey-Fender has worms collected as far back as 1929 among the jars and bottles in the laboratory behind her house. In addition to a few specimens of the giant Oregon worm, she has about eighty species that are still undescribed. As she works her way through the collection, categorizing and describing the native Oregon worms, she publishes her findings in earthworm journals like Reynolds's *Megadrilogica*. "So much work remains to be done," she said. "I have a lot of it well on its way, but I'm no spring chicken."

Her son, William, has taken up the cause, responding to a request a couple years ago by the U.S. Fish and Wildlife Service to go looking for the giant worm one last time. Although other native species turned up, not a single giant Oregon worm was found, even when Fender searched the areas where he'd seen it before. He told a Portland newspaper, "I think we lost this one. I think if there are some left, they won't be around very long."

Still, his mother has vivid memories of the times in the 1980s when she and her husband encountered the two-foot-long white worms. Kenneth once grabbed one, and it wriggled and contracted, using every one of its tiny white hairs to hold itself in its burrow. When it had had enough of being handled, it ejected a good-sized spray of lily-scented coelomic fluid in her direction.

"It was quite an experience," she told a newspaper reporter. In spite of their reputation for squirting liquid when stressed, she remembers a story of a child managing to get ahold of one and swinging it around his head, stretching it (temporarily) to three feet. "It'll do that," she said knowingly.

Sightings of the great white worm are rare in part because they surface only during wet weather and they frequently burrow up to fifteen feet below ground. "When you start to dig," she told me, "they sound. They go down even deeper. Some people use chemicals to get the worms to surface, but we don't do that. We just dig. Sometimes you can shake a post that's stuck in the ground and the vibrations will get the worms to surface, but not these. They don't respond to that."

Most people have never seen the giant Oregon earthworm and would just as soon keep it that way. I asked Dorothy why anyone should care that this worm is nearing extinction. Does it have an extraordinary ability to improve the soil? Does it have a special relationship with any particular plant? She reminded me that given the worm's near-extinction, it is impossible to know. But besides that, she said, "Humans are curious. Humans want to know. We wouldn't be the creatures we are if we didn't. We've got a unique fauna of worms here in the Pacific Northwest, and they are incompletely known. I'm interested in the whole picture. I want to know what's out there." Like any good ecologist, she wonders about the connections between the earthworms and the soil they live in, the forests and prairies they support, the birds or mammals that may have once sought them out for food. Anytime a thread is broken, the web changes forever.

ONE OF THE LARGEST worms in the world is the giant, three-foot Gippsland earthworm, *Megascolides australis,* well known among residents of southeastern Victoria, not far from Melbourne, Australia. Although few people have seen the worm in its natural environment, many have heard a deep gurgling sound from underground, the sound the worm makes when it is disturbed and ducks for cover in deep burrows, sliding through tunnels lined with its own coelomic fluid. Farmers in the Bass River valley may complain that the local conservation movement has gone overboard when it tries to protect its habitat and places the needs of worms over those of people, but just as many will brag about the number of giant worms on their property. Over the last several years, the town of Korumburra has even hosted a worm festival called Karmai, the aboriginal word for giant worm, and crowned an earthworm queen. (Sadly, the festival has been canceled after twenty years in operation.) In spite of all the talk about giant worms in the region, tourists driving along the Bass Highway are often startled to see a three-hundred-foot worm along the side of the road. It is the Giant Worm Museum, the world's only annelid-oriented tourist attraction.

The owner, John Matthews, hadn't taken the slightest interest in worms before the night in 1981 when the subject of the giant worm came up at dinner. The worm had been named back in 1878, and news of its legendary size had even made its way to Charles Darwin while he was completing *The Formation of Vegetable Mould.* Since then, other species around the world were often named after *Megascolides australis,* but in fact, the Australian worm is unique not only in its size, but also in its physiology.

The adults grow over three feet long, although they easily stretch to ten feet. It is almost impossible to handle one of them —their skin is so fragile that it may burst while the worm tries to get away—but a few pictures still exist of people holding one up, three pairs of hands supporting its long pink body.

(Imagine three men holding a ten-foot-long hot dog and you've got the picture.) The cocoons incubate for over a year and when the young worms emerge, they are about two inches long. Although it is impossible to tag them and track their life span, the best guess is that they take up to five years to reach maturity and live twenty to thirty years. It is known for certain that they can survive ten years in captivity. Unlike many worms that heal quickly after a cut and regrow lost segments, the giant Gippsland worm is quite vulnerable to injury. Cut one of these worms and it can bleed to death.

Matthews was intrigued when he heard about them. At the time he owned a hotel and restaurant on Phillips Island, in the Gippsland region, an area already known for the parade of penguins that walk across the beach at sunset. He was interested in opening a tourist attraction. At talks he gave to members of the tourism industry, the subject of the worm often came up. Surely, he figured, this worm was a rarity even among Australia's many natural wonders. The region was already a popular tourist destination—visitors had been flocking to the penguin colony on Phillips Island since the 1920s, and a conservatory allowed visitors to see koalas in their natural habitat—so an attraction centered around the unusual underground dweller seemed like a natural next step.

"We instigated a little research at the university," he said. "There wasn't much data about the worms at first. And we did manage to dig one or two up." When the museum first opened in 1985, only a few live specimens were put on display. A special exhibit even allowed visitors to handle them. "We quickly realized that the worms couldn't tolerate that," he said. "Besides, we couldn't just go dig up more anytime we wanted. Once the museum was built, we drew attention to the fact that their habitat was disappearing. That's when they went on the endangered list." Now visitors have to content themselves with a glass exhibit case in which one or two build deep burrows. There's also a movie theater and eighteen other worm-related exhibits. Because worms are not likely to breed in captivity, it is only a matter of time before visitors won't be able to see a live giant Gippsland worm at all. To hold the public's interest, the museum has recently branched out to include exhibits on koalas, dingos, emus, and other wildlife.

About the museum's architecture, John Matthews told me, "I wanted the entire museum to be in the shape of an actual worm, but it proved too difficult. With so many people coming through, we would have had to build doors and windows along the length of the worm and then it just wouldn't look like a worm anymore. So most of the exhibits are in the building behind the worm structure that you see from the highway. But there is one exhibit inside the worm itself. It's built to look like the inside of a worm body. It's quite realistic. We even recorded sounds of them eating. And we've got worm tunnels you can crawl through to give you an idea of what it's like—you know—what it's like to be a worm."

No one knows how many of these giant worms still live in the fields and river banks of south and west Gippsland. Our very curiosity could be fatal; there is no way to count them without harming them. The giant Australian earthworm lives its entire life belowground in a deep network of tunnels. It deposits its casts where they cannot be seen or measured. Its cocoons hatch underground. It finds its food source among the roots of plants. If it rises to the top of the soil, poking its head out of a burrow to observe the unfamiliar life on the surface of the planet, it is a rare occasion, perhaps not much more than an accident or a wrong turn.

The soil's pressure belowground is tremendous. Pressure increases by 7,300 pounds per square inch for every mile you travel below the earth's surface. It seems improbable that a creature as fragile as the Gippsland earthworm could survive under any pressure at all, but it does. I can't imagine how it digs a chamber for itself underground, a labyrinth of burrows that may extend far below the reach of tree roots, entirely out of our grasp.

When I stand at the edge of a forest, at the base of a mountain, or in my own backyard, looking down at the soil, I feel the way I do when I look out at the ocean, where great blue whales and giant squid swim the unknown depths, where sharks hunt and sea cucumbers wave with the currents. Any sea creature, at any time, can break the surface of the ocean, can rise up from the hidden underwater world and fix one dark eye upon you, then dive down again. The ground has its own kind of fluidity, its own hidden world, its own mysterious inhabitants. What creature, I wonder, would rise up from the surface of the earth if I stood here long enough and watched? How much of the underground world of the giant earthworm is still unexplored and unknown?

# Nature's Plough

One often reads of the thousands of slaves that built
the Pyramids for Pharaoh. In actual fact, these enormous
edifices owe their existence in the main to the thousands
of slaves inhabiting the sub-soil of Egypt.

—ANDRE VOISIN, *Better Grassland Sward,* 1960

NOW THAT I'VE BEEN growing my own vegetables for
several years, it no longer seems remarkable that some of the
food I eat comes from my own backyard. I walk the vegetable
garden every day, and when I see a few asparagus spears, a young,
tightly closed artichoke, or a handful of green pods bulging with
peas, this is hardly a miracle to me. I make a mental note that
we might have an asparagus omelet for breakfast the next morning, but beyond that, I don't think much of it.

It has taken time for the act of growing my own food to seem
so ordinary, so natural. When friends come to visit, they are as-
tonished to realize that I can assemble the better part of a meal
from my backyard flora. It is not difficult to find volunteers to
do my harvesting for me. "Can I pick the onion?" my friend An-
nette asks when she is here for dinner. "Can I dig up a potato?"

Most of us are so far removed from the place where our food

is grown that it can cause a sensation to see it growing, to see groceries on the vine. Picking potatoes out of the loose soil is like pulling Easter eggs out of hiding; they are just below the surface, covered in dirt but still surprisingly bright in hues of pink, yellow, and blue. Brussels sprouts grow on a tall stalk, the miniature cabbages clustered along the length of it. "I don't know how I pictured a Brussels sprout growing," a friend said to me once, "but I never would have imagined this."

If we're so removed from the plants and animals that eventually end up on the grocer's shelf, we are even further removed from the soil where that food has once lived. When I buy an onion at the grocery store, I have no idea about the texture, the color, the composition of the soil where it grew. I don't know what kind of bacteria or fungi fed the bulb as it matured. And I haven't a clue about the earthworm population that might have lived among its roots.

Why does it matter? Because a fruit or vegetable—an orange, a head of broccoli, a carrot—is a product of its environment. It derives its vitamins and minerals from the sun, the rain, and the soil. We expect an orange to provide about seventy milligrams of vitamin C, but does every orange contain that amount? In fact, the vitamin and mineral content of produce can vary widely, and vegetables grown in soil that has been amended with composted manure may provide more nutrition than those fed synthetic chemicals. In one study, conventionally grown beans had only one-tenth the iron of organic beans, and conventional spinach had half the calcium of organic spinach. I don't know of any study that has linked the calcium level in spinach to the number of earthworms in the soil, but I can only guess that those organically grown vegetables had more earthworms toiling at their roots.

This, then, is what is still remarkable to me about growing my own vegetables: I feed the worms in the form of kitchen scraps going into my worm bin, and in the form of manure and dead leaves that I spread around the garden as mulch. And in return, the worms feed me.

I wondered about what earthworms could do for farmers on a large scale. Over the summer, I went to Ohio for the first time. I had come expecting to see fields of young sweet corn, sunflowers, soybeans, maybe a roadside stand, where I could buy some fresh produce. But my strongest memory of that drive through Ohio was of the soil: the deep black earth, impossibly dark, unlike the brown clay fields in California. Here was a place where life had been organized around the soil, where the earth itself was the richest bounty. Occasionally I passed a tractor on the road, and from time to time I saw a farmer riding through the fields on a sprayer, a trail of white mist behind the vehicle. Sometimes I saw nothing but good black dirt for miles.

How often do these farmers stop to consider the contributions that earthworms made to the fertility of their soil? Some of them may have heard about Darwin's assertions that earthworms act as a tiny, but very powerful, plough. A few may have heard about discoveries that soil ecologists are making in this microscopic world. But what would farmers, and, for that matter, gardeners, do with this information? Can growers and plant enthusiasts harness the power of the earthworm for their own gain? As a practical matter, what place does the earthworm have in modern agriculture?

I was about to find out. I'd come to Ohio to meet Clive Edwards, an oligochaetologist who teaches at Ohio State University, a surprisingly elegant old campus established in 1870. I met him at his office in the botany and zoology building, where he shares a crowded room with several graduate students. Color graphs illustrating the results of his latest trials covered the table. A stack of articles awaiting peer review sat nearby. Unlike Sam James's laboratory, there was not a worm in sight.

"I'd take you to the greenhouse," he said, "but we've just finished with our petunias and there's not much to see right now."

The petunia study is one in a series of investigations into the value of earthworm castings for plant growth. "Bedding plants are big business around here," Clive said. "And greenhouse space is expensive. The faster you can grow plants and ship them out, the better."

He leaned back in his chair and described the project. To set up the trials, he began with the standard soilless medium, typically a bark or peat mixture, that most growers use to start young seedlings. Then he started mixing in vermicompost (castings with some compost mixed in), at first adding only ten percent vermicompost to the mix and working up to one hundred percent. He compared those plants to plants grown in the soilless medium only, and to those grown in the same medium with nitrogen, potassium, and phosphorus added to match the level of those nutrients in the vermicompost.

"You don't want to grow plants in vermicompost only," he told me. "It's like any other manure—it's too much on its own. We found that adding twenty percent vermicompost to the soilless medium gets the best results. Look at this," he said, sliding a bar chart over to me. "It may take four or five weeks to get a bedding plant like a petunia grown and out the door. We've been able to cut that time by one or two weeks, and the plants look better."

Although I had no trouble understanding what he meant when he said that the plants look better, as a scientist he wanted a way to quantify that outcome. "One of my students did a very informal study," he said. "He took some tomatoes and marigolds grown the ordinary way, and some grown in twenty percent vermicompost. He set up a booth outside one of our football games and asked people to assess their growth. The plants grown in vermicompost were much preferred over the others. They really did look better. That matters to a grower."

I asked him about crops grown in the field. What happens when, say, a cornfield is amended with vermicompost?

"It's harder to measure," he said. "Too many variables. And if we wanted to incorporate twenty percent vermicompost into an acre of soil, that would mean bringing in twenty tons. It's just not cost-effective. So we tried to determine the least amount of worm castings that would make a difference. We found that two to four tons per acre increased yields, and it also reduced crop loss due to disease and damaged fruit. Not only that, but there was a residual benefit that lasted up to four years.

"Look." He waved his arm over a stack of research findings. "I never set out to prove that organic farming is the only way to go. I'm just trying to figure out how to improve yields. A few years ago, I did a study comparing the yields of conventional agriculture, organic agriculture, and what we call 'integrated low input' agriculture, which uses a small amount of chemical inputs. In the first year of the study, conventional agriculture had the highest, most cost-efficient yields. In the second and third years, the integrated low input farm did the best. And after that, the organic and the integrated low input farms were tied for best economics — even if you don't count the higher prices that organic farmers can sometimes charge for their produce."

THERE ARE PLENTY of stories out there about the ways in which worms can help the farmer. New Zealand, because of its rather limited native earthworm population, has become a case study in the difference that worms can make. Experiments with that country's worm population began not long after Charles Darwin's book was published. Although Darwin focused almost exclusively on the nightcrawler as the protagonist for his book, by the late 1800s, dozens of other species had been identified around the world. "Earth-worms are found in all parts of the world," Darwin wrote, "and some of the genera have an enormous range. They inhabit the most isolated islands; they abound in Iceland, and are known to exist in the West Indies, St. Helena, Madagascar, New Caledonia and Tahiti. . . . How they reached such isolated islands is at present quite unknown."

Even less was known about the difference in function between various earthworm species in the soil, but European settlers traveling to new continents generally considered European worms—if they considered the matter at all—to be best suited for agricultural soil. When those farmers began arriving in New Zealand in large numbers in the nineteenth century, they took note of the native earthworm population but concluded that their own worms would do a better job improving the soil. The deliberate introduction of European worms transformed New

Zealand farmland: when field worms were inoculated into pastures in a ten-meter grid pattern at the rate of just twenty-five live worms per inoculation, the healthy green growth aboveground precisely matched the pattern of worms introduced belowground. The productivity of farmland and pasture increased by as much as seventy percent in the first few years following the introduction of worms. Soil ecologists around the world reported on this remarkable transformation, but New Zealand farmers didn't need a scientific journal to tell them what they already knew: European earthworms could accomplish what no other farming technique ever had. Worms could improve the yield of their crops tremendously.

This fact was not lost on one New Zealand farmer who began transplanting worms into his isolated hillsides in 1925. He dug up chunks of earth from valleys that were already inhabited by earthworms and transplanted those sections of dirt into his own land, introducing them into his soil without the labor—or the shock to the worms—of picking them out of the ground individually. He continued this work into the 1940s, and by 1949, soil researchers visited his farm to document the results. Pastures that were thoroughly populated by European worms grew twenty times more ryegrass and produced more grass for livestock to graze than those that had been left untouched. Thanks to the worms, his flock of breeding ewes doubled in number and he clipped another four thousand pounds of wool in winter. The worms fed the grass, the grass fed the ewes, and the ewes fed the farmer, who in turn fed the worms. It was an extraordinarily successful experiment for everyone involved.

. . .

CLIVE EDWARDS'S RESEARCH got me wondering about the relevance of those New Zealand experiments. Lately I've noticed that some garden catalogs sell worms that you can add to your soil. Often the worms are shipped as infants or not-yet-hatched cocoons, that can, according to the instructions, be easily raked into the soil. I've called these companies and asked what kind of worms they are offering for sale; often they are the very epigeic worms that would be sold for worm bins. Yet these worms wouldn't survive in garden soil unless they were placed under a thick layer of mulch that was continually added to. In fact, it is surprisingly difficult to introduce new worms to the garden: only a few species of earthworms have been successfully raised in captivity in large enough quantities to sell commercially. Epigeic worms like *Eisenia fetida* can be easily raised on a worm farm, but a large anecic worm like *Lumbricus terrestris* doesn't grow well in a closed system and must usually be collected from pastures and lawns if it is to be sold as bait or distributed to farmers to introduce into their soil.

Although these common worms have found their way into some forests almost by accident, it is not always so easy to get a new colony of worms established in a farm or garden. Most experts will tell you that introducing worms into the soil is not worth the trouble, especially given the limited variety of species that are commercially available. The New Zealand technique of transplanting a section of sod from an area high in earthworms seems to work well. But the best method of all is to create ideal conditions for the earthworm population that is already present to flourish—in other words, to feed your worms. At the very least, it would·be a good idea to stop harming them.

I watched the farmers driving sprayers through their fields in Ohio. Even though we call it "conventional agriculture" when a farmer sprays fields with pesticides, herbicides, and chemical fertilizers, these practices have only been widespread since World War II, when the same petroleum-based chemicals used for ammunition and nerve gas were later developed into agricultural fertilizers and pesticides. Farmers practiced organic farming for centuries before that, and it is within this context that the role of earthworms can be best understood.

The case for returning to organic farming goes something like this: farmers have, over these last several decades, taken the multivitamin-and-antibiotic approach to agriculture. When people take a multivitamin instead of eating fresh fruits and vegetables, they lose out on the nutritional value offered by the whole food. The fiber and juice in an orange, the beneficial oils in a walnut, the micronutrients in a spinach leaf are simply not readily available in pill form. In our fast-food society, a multivitamin is a poor substitute for a healthy diet. The same is true of chemical fertilizers in agriculture. In fact, there is a saying among organic gardeners: chemical fertilizers may feed the plant, but organic fertilizers feed the soil. More specifically, when farmers add amendments like manure, kelp, alfalfa meal, and vermicompost, what they are really doing is providing a food source for bacteria, fungi, and protozoa in the soil. Those microscopic creatures, in turn, feed the plants.

There's a parallel between human antibiotics and pesticides as well. As people rely on antibiotics rather than their own immune system to fight off infections, bacteria grow resistant to the antibiotic, and newer, stronger medicines are required to fight the newly resistant strains of bacteria. Likewise, over the last several decades, farmers have relied on pesticides, herbicides, and fungicides to rid their fields of crop-damaging pests, weeds, and diseases. Resistance to chemicals develops quickly, and farmers look to a new wave of chemical treatments for help. Meanwhile, they've killed off some of the living organisms in their soil and beneficial insects in their fields. (In fact, high-nitrogen chemical fertilizers can be so deadly to earthworms that they have been recommended as a way to rid golf courses of nightcrawlers while greening up the lawn at the same time.) This growing dependence on chemicals concerns some farmers, who worry about their increasing cost. It does not sit well with many farmers that their newly sprayed fields are toxic to their children. Add in concerns that many agricultural chemicals are petroleum-based and draw on a limited supply of fossil fuels, and the case against conventional agriculture seems pretty persuasive.

What's even more troublesome is that the disappearance of microbes from the soil means that more water is needed to irrigate the same area, because soil low in organic matter retains less moisture. Runoff from fields finds its way into lakes and streams, where it can harm fish and pollute drinking water. And as the health of the soil declines, the nutritional value of the produce grown in the fields declines as well.

Many of the techniques practiced by organic farmers are designed to support a healthy earthworm and insect population. A border of flowers at the edge of a field attracts bees and ladybugs. A thick layer of mulch holds in soil moisture and feeds the creatures living underground. A scattering of lime on the ground helps correct acid soils and travels into the deeper layers of earth, thanks to the action of worms. Crops are rotated through the fields so that soil-borne diseases don't accumulate in the ground. Every few years, or perhaps at the end of each season, a cover crop is planted to stabilize the soil and protect it from winter erosion. In spring, the crop is chopped down or tilled under and allowed to decompose. Clover in particular is popular because it adds nitrogen and is especially attractive to earthworms. In fact, worm populations are exceptionally high in pastures that are permanently planted in clover, an idea that has caught on among fruit growers in California. Citrus groves, almond groves, and vineyards can all be planted with a permanent undercrop of clover that will attract bees, keep the soil from drying out, and—eventually—decompose into earthworm food. In some ways, it seems like every organic farmer is a worm farmer. Even if you're growing sweet corn or zinnias, hardly a day goes by when you don't give some consideration to the raising and feeding of earthworms.

That's why Clive Edwards's research interested me so much. But before I left Ohio, I ended up grilling him for gardening advice. In particular, I wanted to know what he thought about the debate among organic gardeners and farmers about the best way to prepare the ground for planting. Ploughing farmland—or double-digging garden beds—is a springtime tradition for some farmers and gardeners, but deep ploughing will discourage earthworms and disrupt the soil structure. There's nothing a worm hates more than disturbed soil. Whether it's a farmer tilling hundreds of acres of land each spring, a gardener digging a vegetable bed, or a new subdivision under construction, disturb the soil and you're guaranteed to cut down your worm population.

There's a method of agriculture called "no-till," in which a hole is drilled into the soil just large enough for the seedling or the seed to be dropped in. This method leaves the soil structure intact. It encourages earthworms. Time and again, earthworm scientists have proven how well no-till works to increase worm populations. It's a pretty simple science: they go out to fields and pastures and count worms. On land that has not been tilled, they find more worms and better soil structure every time. Combine no-till practices with consistent use of cover crops, and studies have shown that yields increase by up to sixty percent.

Clive Edwards was adamant about the value of no-till. "There's absolutely no need for deep ploughing of fields anymore. Every year more farmers are switching to no-till. Same goes for home gardeners. People aren't still double-digging, are they?"

I didn't want to admit that as recently as a year ago when I moved to Eureka, I'd had my whole garden rototilled and the entire surface of the land surrounding the house—the grass, the weeds, even a few shrubs—dug out and carted off so that I would be left with an expanse of fresh garden soil, ready for planting. The landscape crew I hired spread around a layer of compost and topped it with chipped bark. Not one blade of grass, not one dandelion, not one blackberry vine remained. It was all carted off so I could have a blank slate where I could be-gin my garden. The landscaper, a resourceful fellow, put an ad in the paper for clean fill and managed to sell the stuff to some-body just down the street who was putting in a retaining wall and needed to fill in some gaps behind the wall. So I saved on dump fees and somebody else got to make use of my weeds and clods of dirt and the worms inhabiting them.

Now I know that there are ways to build a garden without disrupting the earthworm population. Backyard gardeners have been experimenting for years with a layering method that involves chopping the weeds back, spreading cardboard and newspaper on the ground in a thick layer, topping it off with several inches of compost, and seeding in a cover crop like annual rye or hairy vetch. The combination of the cardboard, the compost, and the cover crops smothers the weeds and loosens the soil. In early spring, the cover crops get cut down with a string trimmer. The roots and leaves decompose, the worms work their way through it, and the garden is ready for planting.

I once heard a farmer say this: "The problem with the way we farm now is that we treat our soil like dirt." Indeed, the soil—and the earthworm that lives in it—is itself a crop, one that can be cultivated in the same way that a radish or a tulip is cultivated: by paying careful attention to its particular needs and habits. More and more, I was starting to believe that earthworms may, in fact, be the most important crop I grow.

IF YOU'RE IN DOUBT about the benefits that earthworms can offer to the soil, there's an experiment you can do. It's called a pot test—it's a test that scientists and earthworm enthusiasts have conducted since Darwin's day, and perhaps even earlier than that. White clover is often used in these experiments because it is believed that earthworms have a special relationship with white clover: the worms eat dead clover residues, and the clover flourishes in ground worked by earthworms.

Here's how the experiment works: fill a pot with worm-free earth, and plant clover seed. Fill a second pot with earth, add worms, then plant clover. In a third pot, use earth and dead worms. (How you kill the worms is up to you.) The reason for putting dead worms in the third pot is to make sure that it is not merely the presence of decaying insect bodies that increases the clover's yield. Dead earthworms simply dissolve into the soil, their tissues excreting nitrogen as they decompose.

It's not hard to guess what the results of this experiment will be, but go ahead and try it, and be sure to label your pots and take photographs as you go. Earthworm textbooks from the fifties and sixties are full of blurry black-and-white photographs of clay pots with hand-lettered signs propped in front of them. No Worms reads one, and a few scraggly clover leaves lay limply on the rim of the pot. Dead Worms reads the next sign, and a few more clover leaves rise hopefully from the soil. Live Worms reads the third. Clover spills out of that pot abundantly, a living advertisement for the services of the hardworking earthworm.

There is a widespread rumor among worm growers that castings can prevent pest infestations. Aphids and whitefly will stay away from plants grown in worm castings, the rumor goes. A tea made from worm castings and water can be sprayed on plants to deter all kinds of pests and diseases. A bottle of this concentrated tea sells for around a dollar an ounce—quite a price for watered-down manure. Are these potions nothing but snake oil? Could worm castings really be used for pest control? I asked Clive Edwards about it.

"Actually, there might be something to that idea," Clive said.

"There's an organic farm here in Ohio. These two brothers—the Spray brothers—have been farming one thousand acres organically for decades. They can outproduce the conventional farmers and they're better able to weather bad periods.

"So my colleagues went out to their farm and took samples from their soil. We also took some soil from a conventional farm next door. Same soil types, everything comparable except the type of farming that had been practiced on the land. They took those soil samples into the greenhouse and grew corn and soybeans in them. Then they deliberately released certain pests and diseases into the greenhouse. Would you believe the plants grown in the Spray brothers' soil were practically disease free? And the corn borers wouldn't touch the corn grown in their dirt."

Corn borers are a particularly nasty pest. They are the larvae of a small, yellowish moth, and in that larval stage they tunnel into the tassel of corn ears. Corn borers are unpleasant creatures to encounter in an ear of fresh corn: if you grow corn in your own garden, you will soon make a habit of chopping off the tassel, where at least one corn borer has probably made its home, before you bring it inside.

It's hard to explain exactly why a pest would avoid a plant grown in the Spray brothers' soil. And it's certainly not true that all corn grown in organic soil is free of corn borers. But it does seem clear that sucking insects, those that feed off the sap of a plant, found something distasteful in plants grown in the Spray brothers' soil. Could the thriving worm population in that soil have had something to do with it? When I asked Clive, he shrugged.

"Sure. It's starting to seem like sucking insects in particular may find something they don't like in the sap of plants grown with vermicompost. It could be that the glassy-winged sharpshooter, the insect that's spreading the phylloxera disease to your vineyards in California, reacts the same way. But there's lots of research still to be done."

"Meanwhile," I said, fishing for a little gardening advice, "it may not be *proven* that vermicompost can prevent whitefly infestations, but it probably couldn't hurt to scoop a little out of my worm composter. . . ."

"Oh, of course! There's no reason not to try it in your own garden and see what happens."

I told him I'd give it a try. There's something about the trial-and-error approach to gardening that appeals to me anyway. What attracts me to plant science is not the research, the seed trials, or the charts and graphs. It's the experimentation, the alchemy, the possibility that something entirely unknown and unexpected can happen. An ordinary poppy can break rank with its companions and unfurl a petal with a jagged edge, a dark stripe, or a novel new color. A handful of runner beans dropped in an unlikely location can sprout overnight and scramble up a fence, defying all expectations. A grape grower can spread worm castings around a vineyard, and who knows? Maybe the glassy-winged sharpshooter will stay away. Maybe phylloxera won't spread through the vineyard. That may sound a little visionary, but hope is one of the essential tools of a farmer or gardener. Hope is what gets a person out in the fields in early spring with a handful of seed—hope, and a certain faith in the promise of the good, dark soil underfoot.

# Counting Worms

In the black mealy compost there was twenty worms in
every shovelful, orange worms and purple worms threaded
like screws. There were wigglers whipping in frenzy like raw
nerves and long night crawlers that glistened like snakes,
worms with swole bands and blister rings, worms
that flattened at the edges.

—ROBERT MORGAN, *This Rock*, 2001

WHEN I GET HOME from a trip, the first thing I do is go
outside and check on the worms. A pet-sitter takes care of our
two cats, but I'm not even sure she knows about the existence
of several thousand worms on the porch. They can easily be left
alone for a week or two, as long as the temperatures are not expected to reach too far below freezing or too much above eighty
degrees. The temperatures here hover in the fifties and sixties
year-round. It's the perfect climate for raising worms.

Before I leave for vacation, I add an especially thick layer of
shredded newspaper to the bin, which keeps the bin from getting
too wet, discourages fruit flies, and provides a little extra warmth
if it does get cold. It's not a good idea to add extra food since it
might get moldy before the worms can get around to eating it.
Besides, they can always eat the newspaper if they get hungry.

I came home from Ohio to find that the worms had nearly filled the top tray with castings. It was time to rotate the trays and spread the castings around the garden. I had Clive Edwards's advice in mind; I planned to pile the castings up around the base of some old rosebushes that I'd inherited from the house's former owner. I'm not much good with roses—I'd neglected them, and over the summer they'd become infested with whitefly.

I set the lid of the worm bin on the porch and pulled off the top tray. The second tray was filled with dark, spongy castings and lots of worms. A few melon seeds had sprouted and the young stems, blanched white for lack of sunlight, wove through the castings. I pulled those seedlings and tossed them onto my backyard compost pile with the rest of the garden waste, then lifted the middle tray off the bin and set it down as well, so that I could have access to the bottom tray. The worms had completely eaten through the food, but dozens of them still lived down there in their castings. I suspect that they prefer the bottom tray for laying eggs because it is so rarely disturbed; I can't think of any other reason why they'd still be there. I lifted that tray off too. Below that was the bottom level, where water is supposed to collect until I drain it out by the spigot at the base. A few worms were slithering up the sides of the base, but I didn't bother rescuing them. I figured that if they could climb down there, they could climb back up again. It was time to reassemble the bin.

The middle tray, where the food was nearly all consumed, went on the bottom. The top tray, where I'd been adding food most recently, became the new middle tray. And the bottom tray, with its payload of castings and a few dozen wayward worms, went temporarily on top. Before I could remove the castings, I'd have to get the worms out.

Usually I start by working through the castings with my spading fork to see how many worms I'm dealing with. There might be a dozen; there might be fifty or sixty. Often I turn up worm cocoons this way, too; I try to pick those out and drop them into the tray below, where I hope they'll hatch safely. The worms themselves start moving into the middle tray as soon as I disturb their habitat. I leave the lid off the bin for a couple hours, which lets sunlight in, and I turn the castings occasionally to prod the worms into moving on. Within a few hours, most of the worms have left the tray and the castings are ready to use.

I used to let my castings dry out before I used them, but I found that they formed a hard crust when I did that and eventually turned into light, brittle chunks, like volcanic rock. I've discovered that it works best to use the castings right away, adding them to the soil while they're damp. If I have anything to plant, I put castings in the bottom of the hole. Otherwise, I look around for plants in the garden that seem like they could use a boost and work the castings into the ground with a hand fork. This time I fed them to my roses in hopes that the whitefly problem would clear up. It's never easy to dig anything into an established bed, but I did what I could, loosening the soil around the base of each rosebush and heaping worm castings around the roots.

Once I'd emptied the tray of castings, I quickly set it back on the bin and put the lid on top. I wanted to get back over to the roses. I had visions of whiteflies lifting off the bushes in one winged mass and vanishing from my garden forever. It didn't happen that way, but believe it or not, after a few weeks, I couldn't find a single whitefly in that rose bed. Now, as I write this, almost a year has passed, and there are still no whiteflies.

Next I'll try the same thing with my broccoli and cabbage, which tend to get infested with greyish blue aphids once spring comes on. I can't say for certain that worm castings will be the miracle cure, but I know one thing: I'll keep experimenting until I find out. The worm population in my bin is now seven years along; many generations have lived there and reproduced. They are generating enough castings for all the experiments I'd like to conduct.

I'VE MADE A REAL effort to increase the earthworm population in my garden soil, too. I've brought in compost and mulch, I've planted cover crops and spread lime around to cut the acidity. After a while, I was ready to see if all my efforts to attract earthworms had amounted to anything. I waited for a damp, cloudy day when earthworms would be near the surface, and I went outside with a bucket and a shovel. I was going to take an earthworm census.

I picked a vegetable bed where I knew they would be plentiful and the ground would be loose and easy to dig. As soon as I turned over a clump of earth, I saw that my efforts with the soil had been rewarded: four or five huge nightcrawlers appeared on the end of my shovel, and several smaller worms—perhaps *Aporrectodea caliginosa*—ducked away from the light. The soil was perforated with holes about the diameter of a pencil—

burrows. The worms, it seemed, had multiplied and occupied my garden soil in greater numbers than I could have imagined.

A ten-quart bucket sat next to me. I dug out enough dirt—and worms—to fill it. Once it was full, I sat down next to the bucket and started counting worms.

This hand-sorting technique can be time-consuming if it's done on any large scale. There are other methods of sampling earthworm populations: a strong vibration will bring them to the surface, and some chemical extractors will also force them out of their burrows, although worms might not survive the exposure to chemicals. And as Sam James pointed out, wasabi and mustard will also force them to the surface. But I didn't need any extraordinary measures to carry out this small earthworm count. Sorting them by hand would work just fine. In fact, it was a pleasant task, sifting through the bucket, examining the denizens of my garden soil up close. I was looking for worms, but I found quite an assortment of other creatures: ants, pill bugs, tiny spiders, and shiny dark beetles that scurried away before I could get a good look at them. There were cutworms and grubs and slugs, and tiny white specks I thought might be springtails. I held the larger insects in my hand and then set them aside, looking for worms. One thing was certain: it was crowded down there in the ground below my garden.

It took me about a half-hour to sort through the bucket and count the worms. I was careful to break up each clump of dirt to find the worms that might be burrowed inside. There were plenty of baby worms, along with about a half-dozen worm cocoons: sure signs that the worm population was healthy and ready to expand. I turned up close to a dozen six-inch-long nightcrawlers as well. In all, there were forty worms in my bucket.

If I wanted to do a truly scientific study, I would have chosen several different sites to sample: in addition to this fertile vegetable bed, I would have taken worms from the front yard, which is planted in perennials, and from the paths, where I have done little to increase the soil fertility. I might have even taken a sample from the alley behind my house, where the soil has probably not been disturbed for decades.

But I wasn't interested in disrupting the earthworm community any more than I absolutely had to. Besides, I was mostly interested in knowing what kind of potential my soil had—what population size was possible if I brought in compost and manure, planted cover crops, and did everything right.

Most worm scientists describe worm populations in terms of worms per acre. Darwin estimated that there were fifty-three thousand worms per acre, and population studies performed in the twentieth century have shown that earthworm populations can range from as few as twenty thousand worms per acre in Rumania, to six hundred and seventy thousand in tropical rainforests in Malaysia, to an astonishing eight million worms per acre in a New Zealand pasture. And the experience of farmers and ranchers in New Zealand shows that where earthworms are abundant, crop yields increase.

It took a few minutes in front of the computer to convert my ten-quart bucket to cubic feet, and then (assuming that I was only really sampling the top twelve inches of soil in my garden), to calculate the square feet in an acre, and in the end, I calculated that if all the surrounding soil was as rich as the bed I dug, I'd have 5.2 million worms per acre. My house sits on a lot that measures about one-eighth of an acre. At that rate, there would be over six hundred and fifty thousand worms, almost a hundred times more worms than Darwin's highest estimates.

What could this mean for the fertility of the soil in my garden? Estimates of the volume of earthworm castings deposited in the dirt each year vary widely and depend on the available food sources and the conditions—temperature and dampness—of the soil. One conservative estimate is that a single worm can eject a little over an ounce per year. That would work out to about 150 tons per year, per acre, which seemed like an extraordinary figure when I thought about Clive Edwards telling me that it would be too expensive for a farmer to import twenty tons of vermicompost per acre each year. (Of course, one critical difference between his studies and my estimates is that he was using a very high-quality vermicompost product made under controlled conditions, but the earthworm casts ejected by night-crawlers in my own yard may be of a much different—perhaps lower—quality.) Also, I know that earthworms aren't as abundant throughout my garden as they are in one enriched sixteen-square-foot vegetable bed. But even if I consider the fertility of that area alone, it is astonishing to think that the earthworm population of that bed—about two thousand—is capable of producing 125 pounds of castings for my vegetables every year.

Figures like these give scientists a good reason to see earthworms' potential. They lend credence to some primitive notion, some early impulse, that tells us that worms must perform daily miracles underground. I have an old book called *Humble Crea-tures* that was written just after Darwin's first paper for the Ge-ological Society, but before the publication of *The Formation of Vegetable Mould*. The purpose of the book was to introduce read-ers to two humble creatures they might never have thought much about: the earthworm and the housefly. The author, James Samuelson, had a fervent admiration of the earthworm's abili-ties, which was bolstered in part by Darwin's own work. Samuel-son quoted liberally from Darwin's early paper on worms but concluded with his own graceful observation of the earthworm at work: "There it toils away, unconscious of its great mission, again and again penetrating the earth, and each time, when it returns to the surface, bringing up with it a small portion from below to aid in the restoration of the exhausted soil, and multi-ply the comforts of the human race."

SURELY A CRATE of potatoes stored away for winter, a rich sweet orange, or an armload of deeply perfumed roses are among the great comforts of the human race. The earthworm plays some role in making all those comforts available to us. It is easy, I have found, to get a little sentimental when it comes to their contributions. Even a scientist can become consumed with a sense of wonder over the power of the lowly worm.

Since Darwin, one of the most ardent admirers of earth-worms was a French scientist named Andre Voisin, who studied grassland ecology. He grew so enthusiastic about the achieve-ments of earthworms that he published a paper in the early 1960s claiming that worms could be responsible for the development of the world's great civilizations. Although his ideas probably seem a little crazy, I believe they might not be too far from the truth. Earthworms, he discovered, populated the Nile, the Indus, and Euphrates valleys in unusually high numbers. He concluded that great civilizations flourished in those areas in part because a healthy population of earthworms was present to work the soil.

"Civilisations could flourish," he wrote, "in regions where the soil, to use Darwin's words, had passed thousands of times through the intestines of active earthworms." He went on to make what is perhaps the most surprising argument to date in favor of the accomplishments of earthworms: "Many members of the human community, instead of cultivating this land that had already been so well ploughed, were going to devote their time to the construction of buildings and to creating works of the mind."

Voisin was not the first person to suggest that great civilizations spring up when the conditions are right, as if people are like gardens, blooming and expanding only when the soil is rich, the rain plentiful, and the climate temperate. He was influenced by Ellsworth Huntington, a Yale professor of geography, who wrote about the impact of climate and geography on the development of human civilization in his book, *Mainsprings of Civilization,* published in 1945. He drew maps of the planet on which zones of "human progress" were superimposed over zones of favorable climactic and geographic activity. The Nile, Indus, and Euphrates valleys ranked highest on his map of human progress. Fertile soil, he seemed to suggest, is a powerful agent for change. It feeds people, and once they're fed, they can go about building a new society for themselves. People can live closer together and share the labor of farming, even creating specialized jobs (blacksmith, baker, carpenter) and, eventually, developing tools and technology to make those jobs more efficient.

If all of human progress springs from good soil and warm summers, shouldn't earthworms get some of the credit? Voisin thought so. Thanks to them and their ability to plough the soil, he reasoned, people were, for the first time, free to devote their spare hours to the development of math, science, a written language, and even the construction of the pyramids in Egypt.

Voisin was a first-rate earthworm romanticist, lavishing praise and poetry upon them. I can't help but admire his vision, his faith in what an earthworm could do. One of the great snippets of earthworm poetry comes from Voisin, when, in the middle of an otherwise ordinary chapter on earthworms and agriculture, he conjured up an image of Hamlet and Horatio standing on the terrace at Elsinore and discussing the lowly earthworm's achievements:

> [T]his evening there is no mention of the ghost that walks the terrace by night. The light of the moon suddenly falls upon an earthworm sliding under a stone, and Hamlet asks his friend:
> "Did you know, Horatio, that without earthworms men could not create civilisations?"
> With characteristic scorn Horatio answers sarcastically:
> "Until now I thought that earthworms were destined to

destroy the last traces of human civilisation, devouring men's corpses and swallowing up their buildings."

To which Hamlet replies once more:

*There are more things in heaven and earth, Horatio,*
*Than are dreamt of in your philosophy.*

# Garbage into Gold

When stepped on, the worm curls up. That is a clever thing
to do. Thus it reduces its chances of being stepped on again.
In the language of morality: humility.

—Friedrich Nietzsche,
*Twilight of the Idols*, 1889

After spending so much time with earthworms, I
have developed an appreciation for their essential qualities.
Worms are ruminators; they sift through whatever surrounds
them, turn it over, explore it, move through it. They are deliberate creatures, in no great hurry, but always in motion, twisting
and burrowing, shrinking and contracting, and eating. They
spend their lives in a kind of active meditation, working through
the detritus in which they live, the bits of leaves and grass and
particles of soil. For a being with such a simple brain, a worm
seems, in this way, almost thoughtful.

Functionally, worms really do only one thing: they digest.
They live in their food source and their own waste is not repul-
sive to them; in fact, the bacteria in earthworm castings help to
build the kind of soil community where they can thrive. I suppose any kind of digestion is transformative: any food source

that is eaten becomes something else. Any environment, any single life is in a continuous state of change. This is just more obvious when you pay attention to earthworms. Their work may seem unspectacular at first. They don't chirp or sing, they don't gallop or soar, they don't hunt or make tools or write books. But they do something just as powerful: they consume, they transform, they change the earth.

Digestive problems plagued Darwin for most of his life. He took to his bed for weeks at a time, unable to write or keep up with his correspondence. His illness often worsened when a manuscript was due to the publisher, or during one of Emma's ten childbirths. In fact, even when Emma was confined to bed during the last month of a pregnancy, she still looked after her husband. When a child became ill, Darwin often required a nurse as well. The man who sailed around the world on the *Beagle*, the great scientist, the father of the theory of evolution, spent most of his life in a sick and fretful state.

He suffered from bouts of vomiting, boils, dizziness, and headaches. He availed himself of every cure that Victorian medicine had to offer him: small doses of arsenic to settle the pain brought on by inflamed lesions; batteries that gave an electrical shock to an upset stomach, ointments and tonics of all kinds, and a water cure that involved alternating cold-water baths and sweats and required him to wear a cold, wet compress on his stomach most of the day. He kept a diary of his illness, assigning codes to each symptom. In 1852, seven years before the publication of *Origin of Species*, he recorded an equal number of sick and healthy days. Is it any surprise, then, that Darwin so valued the vigor and the digestive powers of earthworms?

Psychotherapist Adam Phillips pointed out that Darwin, like the worms, had a great deal of material to digest, and his work, like the work of the worms, would change the face of the earth. Exploring the psychological implications of Darwin's work, Phillips wrote, "When it came to earthworms, suffice it to say, their digestion worked. Indeed digestion was their work, and it had the most remarkable consequences. Work and digestion—and the links between them—were to dominate Darwin's life. For the idea of work as digestion, and digestion as the body's forced and unforced labour, Darwin turned to the worms."

Earthworms, then, may have been the perfect subject for him. They were built only for eating, and they carried out their task flawlessly. They seemed to suffer no illness and rarely faced any obstacle that could keep them from their work. This may explain why Darwin heaped so much praise upon them. In choosing the earthworm for the subject of his final book, he revealed something about the moral universe in which he lived. Phillips wrote, "It was to be part of Darwin's undogmatic shifting of the hierarchies to see earthworms—typically associated with death and corruption and lowliness—as maintaining the earth, sustaining its fertility. The poor, he would imply, had already inherited the earth. . . ." Through his exploration of this overlooked creature, Phillips points out, Darwin wrote his own version of *The Making of the English Working Class*. The great scientist discovered more about earthworms than merely their habits and habitats; he revealed their noblest qualities.

Noble? A worm? I've watched earthworms squirm around in my garbage for years. I suppose I can understand why people associate them with filth; after all, they are so often found writhing beneath the muck and trash. Anything that's rotten will attract them, but they are not rotten themselves. Anything dead will have a worm writhing through it eventually, but worms are powerfully alive. I have come to understand, like Darwin had, that earthworms are not destroyers, but redeemers. They move through waste and decay in their contemplative way, sifting, turning it into something else, something that is better.

SOMETIMES PEOPLE ASK me why I've never started a worm farm. After keeping them for seven years, I guess I've got the knack of it by now. Maybe I could expand my operation, build a few wooden bins in the backyard, and make a little money selling worms to people who want to get started with their own bins. Maybe I could even offer to pick up kitchen waste from a few restaurants in town, charging them less than they pay in disposal fees and selling the castings for a premium price at nurseries. Once, in Pasadena, I saw a quart-sized carton of worm castings selling for fifteen dollars. That sounds like easy money to me. So why not?

For one thing, I'd run into the same problem selling worm castings that I would if I were to go into business selling chocolate or tawny port: I'd keep all the inventory for myself. I buy compost by the truckload for my annual mulching of the flower beds. I have to buy it because my worms only produce a few cubic feet of castings each season. If I had enough worms to generate truckloads of castings, my farm would still have only one customer: me.

Also, worm farming has a bad reputation as a get-rich-quick scheme. Over the years, people have heard about the great potential in earthworm castings and hoped to make a fast buck. After all, you can get started in worm farming with a fairly small investment, and they don't take up nearly as much space as other get-rich-quick farm schemes like llamas or chinchillas. Maybe that's why so many people have staked their fortune on a box of worms. Out here in California, it seems as if everyone over the age of sixty had an uncle or a neighbor who tried to raise them for a while. Worms and rabbits were a popular combination: both reproduce rapidly, promising more profit, and if the rabbit cages are suspended above the worm beds, the worms will eat the droppings along with any alfalfa pellets the rabbits miss. But most of these stories end the same way. The worms don't reproduce as fast as promised, or they dry up in the heat. Sometimes the company that promised to buy back the worms at an agreed-upon price won't live up to its half of the bargain. Eventually the enterprise fails and the uncle or the neighbor moves on to something else.

In fact, the story of the uncle who tried worm farming is so ubiquitous that e.e. cummings even wrote a poem about it. In his poem "nobody loses all the time," he laments the failures of Uncle Sol, who tries to grow vegetables but fails because the chickens eat them all. Then he becomes a chicken farmer, but the skunks eat the chickens. He becomes a skunk farmer but the skunks all catch a cold and die. The poem ends with the family gathered around Uncle Sol's casket. He redeems himself in death, finding success as a farmer as he is lowered into his grave:

> i remember we all cried like the Missouri
> when my Uncle Sol's coffin lurched because

somebody pressed a button
(and down went
my Uncle
Sol
and started a worm farm)

But the real reason I've never become a worm farmer is that even after all this time, I get a little nervous about the idea of keeping them in such mass quantities. The more I know about what they're capable of, the more I wonder whether it is such a good idea for them to outnumber me a million to one right here at home. Worms are fairly simple creatures, but they can be difficult to manage in large numbers. I heard a story about a couple who had gotten into worm farming and awoke one morning to find that their hundred thousand worms had walked out in the middle of the night. The worms were clustered in balls all over the walls of their barn, and there was no way to round them up. It was as if they'd gone on strike to protest the conditions in their bedding. Without moisture or food, they would probably die there. It's a ghastly image, all those dead worms stuck to the walls.

Water is the thing that most often makes worms go on the move like this, as if they are following some evolutionary signal to move away from the sea and seek dry land. Sometimes they will walk out in a rainstorm and collect on the sidewalks to get away from the high humidity in their burrows. There have been reports of worms climbing to the rooftops of buildings in Burma. These walkouts are often fatal—if the sun comes out, a worm will get cooked—but if things get bad enough where they are, they'll take their chances anyway. There's nothing you can do about it; maybe that's what worries me. I like to think that I'm in charge of my worms, but am I?

I've come home from vacation once or twice to find that the lid of my bin has blown off in a storm. At first my heart skips a beat: what if the worms staged a walkout while I was gone? But then I realize that they would not get far on the wood plank surface of my porch. If they left, I'd know it, because the porch would be covered with dead worms. But so far, they have never walked out on me, even when the lid was removed. No matter what shape the worm bin is in, when I lift the layer of newspaper, underneath I see them all, loyal and content. The lid is usually in the backyard. I retrieve it and place it gingerly on top of the bin, as if I am pulling a blanket over the worms to tuck them in for the night. Underneath all my worry about walkouts and the hidden agenda of worms on the move, I do have quite a bit of affection for the creatures. That's another reason why I've never started a worm farm. I hate the thought of packing them up and sending them off to live with strangers, even for good money.

So I stick to my small-scale bin. A few of my friends have them now, too. We compare notes, rummage through each other's bins, and discuss feeding strategies and reproductive rates. I hardly think of a bin as an unusual thing to have on the back porch, but in fact, the idea is fairly new. Thirty years ago, there were very few guidebooks and no commercially manufactured worm bins. All that changed when Mary Appelhof, one of the best-known worm composting advocates, published a brochure on the subject and sold it to anyone who would send a quarter and a self-addressed, stamped envelope. By 1982, that brochure had evolved into *Worms Eat My Garbage,* a book that is now the classic how-to guide for home worm composters.

Appelhof told me that she had no qualms about the idea of millions of worms going to work on garbage. "When I got started thirty years ago, that's what I envisioned—huge piles of garbage and huge quantities of worms. I didn't have the wherewithal to make that happen, but I did know how to get worm composting going one household at a time. So that's what I did.

"I used to say that one ton of worms could eat one ton of garbage. I was always thinking big like that. Then I found out that Seattle had distributed four thousand worm bins. I did some figuring and realized that worked out to ten tons of garbage going into worm bins. That's when I realized—it's happening! It just isn't happening the way I originally thought it would."

Ten tons of garbage. It is staggering to think of the amount of waste that people produce. Californians dispose of about thirty-eight million tons of waste per year. I did the math; it works out to seventy-two tons per minute. I can't get my mind around a figure like that; I can't imagine what that much waste would look like. I think about the handfuls of kitchen scraps I toss to my worms. That's fine, but what about the tons of waste going into the landfill every minute? Couldn't some of that garbage be fed to earthworms too?

I TRY TO PICTURE a wooden bin full of rotting food and earthworms behind an upscale Manhattan restaurant. The kitchen staff would have to separate vegetable scraps from the meat, the grease, and anything that's been seasoned or covered in sauce. Someone would have to carry the food out to the alley and feed it to the worms every night, shining a flashlight down into the bin to make sure there were no problems with mold or excessive dampness. They would have to be protected from the cold in winter and the heat in summer. Most important of all, they would have to be locked up to be kept safe from mischief in the middle of the night. And when the bin was full of castings, someone would have to take the time to harvest them, separate out the worms, and—what? Push them down Madison Avenue in a wheelbarrow?

It's not a likely scenario. But hauling food waste to the landfill gets expensive, not just for restaurants, but for breweries, food processing plants, and cafeterias. Even dairy and livestock farms produce a waste product—manure—that has to be disposed of somehow. But it takes a large-scale facility to make worm composting work in a way that is clean, efficient, and—especially if the sight of millions of worms writhing around in garbage makes you squeamish—out of sight. In the last decade or so, a few companies have started to build massive new worm composters that could handle waste on a grand scale.

Unlike a home bin, these larger models, sometimes called "continuous flow reactors," are highly automated. Many of them look like large metal bins—picture a series of garbage Dumpsters all joined together—with machinery underneath and at either end to apply food and remove castings. A tray powered by a conveyor belt moves along the top and deposits food at the surface, hydraulic scrapers remove castings from underneath the machines, and thermostats monitor temperature constantly, activating fans, sprinklers, or air conditioners whenever the feed-stock gets too hot. These systems, which can cost up to $100,000 to build, are usually housed in simple modular warehouses. They are more like machines than ordinary bins, and the earthworms are like industrial workers, each doing one small job on a massive factory line.

I talked to Scott Subler, who worked with Clive Edwards at Ohio State University for seven years before he founded his own worm compost company. "We were seeing some incredible results with our vermicompost trials at OSU," he told me. "I got so interested in the waste management side of worm composting that I decided to leave academia and form this company. Because I was in Ohio at the time, I could tell that manure management on farms was turning into a major pollution issue. I could have set up a paper waste or a food waste facility, but I really wanted to see what we could do about animal waste."

Subler's not the only one concerned about animal waste. The Environmental Protection Agency has stepped in with increasingly strict regulations governing the management and treatment of animal manure. The agency estimated that five hundred million tons of animal manure are produced by the major dairy and livestock farms every year. That manure can be spread on farmland as fertilizer, or simply heaped onto empty fields as a way to get rid of it. However, the EPA has found that runoff from large animal operations can pollute drinking water supplies with unhealthy levels of nitrate and dangerous microorganisms like cryptosporidium. It can add too much ammonia to streams, rivers, and lakes; contribute to blooms of algae; and deplete the level of oxygen in the water, which can cause fish to die off. Even air quality is affected: these large operations emit a tremendous amount of methane gas and release ozone and greenhouse gases that the can reduce crop yields, make plants more vulnerable to disease, and cause health problems for people working around the farm.

"It's going to cost more and more for farmers to haul manure away," Subler told me. "To clean this waste up to the point where it's not harmful will require something that's almost the equivalent of a sewage treatment plant. It's a big issue. That's why I decided to focus on dairy manure."

His facility uses one of the high-tech continuous flow reactors. It is housed inside a metal warehouse, where staff can control the temperature and constantly monitor the manure coming in and the castings going out. The facility is so clean, the machinery so streamlined, that it's possible to forget that inside the reactor, thousands and thousands of earthworms are eating their way through the feedstock, just as they do on a much smaller scale in my own backyard bin.

"We can process 250 to 350 tons of manure per year," Subler said. "We pre-compost the stuff to kill weed seeds and pathogens, then run it through the reactor. The worms love it and you already know it makes a good soil amendment. And the best part is, we built this reactor with off-the-shelf materials. All the moving parts are the same ones they use for dairy and manure handling equipment already. It's easier for farmers to get parts and make repairs that way."

Once the equipment is set up, the next question is what to do with the castings. Subler has decided to focus his efforts on creating a consistent worm casting product for home gardeners.

Consistency is a major issue for anyone in the compost business; in fact, Clive Edwards is working with a U.S. Department of Agriculture committee charged with setting nationwide quality standards for compost. Edwards and Subler both agree that consumers should know what they're buying when it comes to vermicompost in particular. The quality can vary widely depending on what the worms were fed and the extent to which they ate their way through the foodstock. (When worms are fed composted cow manure, for example, it may be difficult to tell by looking whether the finished product has been completely digested.) As Edwards and his colleagues start to publish their findings about the benefits of vermicompost in agriculture, they want to make sure that farmers and gardeners understand the importance of a consistent product. "We also don't want manufacturers adding fertilizers to the compost unless it's labeled properly," Edwards said. "People think that if a little bit of something is good, a lot must be better. But that's not always the case."

So Subler sells bagged earthworm castings, a liquid "casting tea," and tea bags that gardeners can use to brew their own liquid fertilizer. He's even gone on QVC to sell his product and spread the message about the benefits of worm castings. When I spoke to him, our conversation focused not on boutique products for gardeners, but on the broader implications of earthworm farming and animal waste. After all, fertilizer isn't cheap, and it's no secret that the chemicals farmers apply to their crops are costly, hazardous to the environment, and increasingly unpopular with the public. There are just under one billion acres of farmland in the United States. What if all that livestock manure—five hundred million tons per year of it—could be digested by worms and applied to nearby corn, wheat, or soybean fields? Sure, that only works out to a half-ton of castings per acre per year, which falls short of the two to four tons per acre that Edwards is considering in his research. But what if composted food and paper waste were added to the mix? Is it too idealistic to imagine that worms could take care of all kinds of agricultural, industrial, and municipal waste and turn it into a fertilizer that farmers could use in place of chemicals?

"It's a real policy issue, putting it all together like that," Subler said. "The inputs are not a problem. You're right, there's animal waste, food waste, all kinds of organic material for the worms to eat. It's the output that's a problem. To create a market for worm castings, the farmers have to be encouraged to use vermicompost. The supply has to be consistent and available when they need it. As a concept, it makes sense for everybody to implement it—regulators, business people, farmers, the general public—but it's still hard to make it go. There are maybe a dozen people in the country working on this. But give it another ten years. It'll catch on."

FROM THE WORMS' PERSPECTIVE, large-scale composting must be a strange new world. They've made no conscious attempt to befriend us; instead, they have just continued their slow, incremental work, adapting to the intrusion of our streets and cities the best they can. But now, only recently, we have begun to take notice of them and to realize that they could be of some service to us. We lay this gift in front of them: our garbage. In doing so, we may set about a revolutionary change in the long history of earthworms on the planet. For the first time, we are attempting to domesticate worms.

A species has been chosen—in this case, *Eisenia fetida*—that has some qualities that are useful to us. Isn't that how we always choose an animal for domestication? A dog guards the house, a cow produces milk, a honeybee pollinates the crops and deposits honey into the hives we build for them. So now we're striking a bargain with worms, offering them the same deal we've always offered to creatures that are useful to us: food, lodging, and unchecked reproduction in exchange for their labor, their output.

Darwin had a keen interest in domestication and its effects on species. Here was an example of evolution at high speed—a carrier pigeon, a racehorse, or even a tomato could be bred and cross-bred to suit our purposes, altering the species far more quickly than it would ever evolve in nature. Even without deliberate cross-breeding, he realized that a creature could adapt itself to us if that was in its best interests. If a slight variation in a hen or a housecat makes it more successful around humans—gets it more food or better shelter—wouldn't that variation gradually become more pronounced in each successive generation? I have an unusually sweet-tempered cat that never claws the furniture, keeps his long silky coat free of mats and mud, and stays near me without demanding too much of my attention. This cat happens to display the qualities that I appreciate in a pet. In the long run, a cat like this might reproduce more successfully thanks to its ability to please humans. It is pleasant enough to think of those traits migrating through the feline gene pool over time. But it is also an extraordinary influence for us to have over any creature. In his book *The Variation of Animals and Plants Under Domestication*, Darwin described this process of selection as "the paramount power, whether applied by man to the formation of domestic breeds, or by nature to the production of species."

Something else Darwin wrote stuck with me. He was describing the accidents of breeding that can happen during domestication—the way that selecting an animal for one trait might encourage other related, but not necessarily desirable, traits to prosper as well. He probably wasn't thinking about earthworms at the time, but what he said struck a chord, because he mentioned the one anatomical feature that is indispensable to an earthworm: its intestine. Writing about the power of the clumsy human breeder, he said, "In the living animal or plant he cannot observe internal modifications in the more important organs; nor does he regard them as long as they are compatible with health and life. What does the breeder care about any slight change in the molar teeth of his pigs, or for an additional molar tooth in the dog; or for any change in the intestinal canal or other internal organ?"

The truth is, there is little cause for worry about the future evolution of the earthworm's intestine. Earthworms are simple, stable creatures; there is not a great deal of variation between two worms of the same species anyway and domestication is not likely to give rise to any highly unusual new traits. Let them eat our garbage—we can use their help, and their castings. And there is still a great deal of genetic diversity among earthworms. There are plenty of worm species left in the wild, species that are wholly unsuited for domestication and will continue their work in the soil as long as we don't disturb them or drive them out.

*Eisenia fetida* is the perfect worm for domestication. It thrives in waste. A pile of garbage is its home. The fact that we can also use it to bait a hook or feed a chicken makes it all the more suited for life alongside humans. In fact, unlike the domestication of cats or dogs, the domestication of the red wiggler requires no changes on the part of the worm. It is already fully equipped to do the job we're asking it to do, which seems like an extraordinary coincidence. Perhaps *Eisenia fetida*, by following the spread of humans around the globe over the last few million years, and moving into our paddocks and our garbage dumps, has already adapted itself to us. Maybe it has been domesticated all along, and it has just been waiting for us to notice.

# Public Duties

... its history affords a striking exemplification of the divine
truth, that no creature has been formed without its special
ends, and that the humblest are frequently selected to
carry out the most gigantic natural operations.

—JAMES SAMUELSON, *Humble Creatures:*
*The Earthworm and the Common Housefly,*
*in Eight Letters,* 1858

"WORMS HAVE PLAYED a more important part in the
history of the world than most persons would at first suppose,"
Darwin wrote in the conclusion of *The Formation of Vegetable
Mould.* One biographer reports that his friend and colleague
Joseph Hooker said this about Darwin's book: "I must own I had
always looked on worms as amongst the most helpless and un-
intelligent members of the creation; and am amazed to find that
they have a domestic life and public duties!"

Those public duties extend far beyond the reaches of Darwin's research. Even he might not have imagined that in the
coming century, worms would be shown to have the ability to
transform a forest, to destroy a rice terrace, and to accelerate the

growth of greenhouse plants. He surely did not foresee the invention of a worm "reactor" that would harness the power of earthworms to consume tons of garbage. And he would have been astounded at the notion that anyone would try to make their fortune by raising earthworms for profit.

It is hard to predict where earthworm science will be at the end of the twenty-first century. But some tentative steps have been taken to solicit the assistance of earthworms in yet another human endeavor: the cleanup of pollution and the prevention of further damage to the environment. Since many pollutants eventually make their way into the soil, filtering down into the dark, damp places where earthworms live, this is, ironically, an issue as important to their future as it is to our own.

It was not until the mid-sixties that a connection was made between earthworms and pollution. Worms had a talent that no one had paid much attention to before: they could take up pollutants in enormous quantities and live. Rachel Carson told the story in her groundbreaking book, *Silent Spring*.

Early studies showed that DDT (dichlorodiphenyltrichloroethane), an insecticide used to combat mosquitoes and lice, and thereby help fight malaria and typhus, would not gather in high enough concentrations in the soil or water to harm wildlife. This was proven wrong on several levels, and the practice of spraying DDT was stopped in this country thanks in large part to Carson's book. Its use was widespread in the United States starting in about 1945; in fact, over a billion pounds of the pesticide was sprayed in this country during the thirty years it was in common use. By the early seventies, DDT was banned in the United States, although it is still used around the world, particularly in Africa, Asia, and Latin America.

It was Carson who reported that worms have an amazing ability to absorb whatever is in the soil, and they were able to take up huge concentrations of DDT into their tissue and still live. This was particularly a problem where elm trees were sprayed with DDT. The chemical remained on the leaves, not even washed off by rain, and in the fall, the trees dropped their leaves. Over the winter worms consumed the decomposing leaves, taking in large quantities of DDT. In the spring, the natural life cycle of worms and birds played itself out as usual: the worms came out of the ground at night; the birds foraged for stragglers early in the morning. But this time, the robins that ate worms also ingested what was stored in their tissue—a massive, concentrated dose of DDT.

Eating only eleven of these worms would be enough to kill a robin, and when you consider that a robin can eat ten to twelve worms in an hour, it's no surprise that the worms were toxic enough to kill them. Some robins survived the high doses of poisoned earthworms, but were infertile and did not lay eggs the following spring. The lesson here is that earthworms were never factored into the analysis—such as it was—of the possible consequences of spraying DDT. Worms were simply overlooked, ignored, misunderstood—and their impact, their importance to the soil and to the ecosystem as a whole—was far greater than anyone could have guessed.

The fact that earthworms absorbed DDT in such high concentrations has not been lost on environmental scientists working in the areas of land reclamation and toxic clean-up. Perhaps the most significant achievement of earthworms yet will be their ability to help us out of some of the most awful messes we've created.

Earthworms have been used as biomonitors at toxic waste sites, where they quickly take up pollutants into their bodily tissue (they are particularly known for their ability to absorb metals like lead), often surviving long enough to be collected by monitors and tested. In this way, worms become the canary in the coal mine, giving a clear picture of the extent to which chemicals present in the soil or groundwater are affecting soil organisms and, by extension, other animals living at the site. Over the years, protocols have been developed for introducing earthworm species into contaminated areas and testing them for exposure to pollution. The EPA, the U.S. Army Corps of Engineers, and other agencies have developed programs using earthworms as biomonitors. These tests, called "bioassays," are particularly important for monitoring the many potential pollutants for which there are no other conventional measurements. By tracking the levels of toxins in earthworm tissue, scientists can now monitor pollution issues that were previously difficult to quantify, such as the cumulative effect of pollutants over time and the impact of several toxins in combination.

Some of the most exciting new work with earthworms goes beyond using them simply to monitor pollution. Scientists are now exploring ways in which they can be used to break down toxins and actually clean up pollution. Dr. Andrew Singer at Oxford University recently investigated the use of worms to clean up soil contaminated by polychlorinated biphenyl (PCB).

PCBs are mixtures of synthetic organic chemicals used in electrical equipment, paints, rubbers, dyes, plastics, and many other industrial applications. The EPA banned manufacture of PCBs in the United States in 1979 and has been overseeing a gradual phase-out of their use in existing equipment, as well as clean-up of sites contaminated by PCBs, since then.

"As it turns out, PCBs were first detected in bird extracts when scientists were looking for DDT," Singer told me. "They didn't know what the PCBs were at first. They just thought they were some weird anomalies— this is back in the mid-1960s— then a researcher just got lucky and figured out that the anomalous chemicals that were being extracted from birds all over the place were in fact PCBs, and thus the explosion of interest in PCBs in the early 1970s. Somewhat similar to the DDT story."

PCBs have been described by the EPA as "extraordinarily toxic." It has been estimated that there are over one billion pounds of PCB compounds in the environment. They can accumulate in fatty tissue and have shown up in the breast milk of Inuit women in northern Quebec. Although PCBs have been identified as possible carcinogens, their worst effects seem to be related to the nervous system and brain functions. Birds who have high accumulations of PCBs in their tissue end up with odd genetic defects and are sometimes unable to build a proper nest. They are dangerous, persistent toxins that do not easily break down.

Singer knew that some work had been done on the use of bacteria to break down PCB in the soil. However, it was proving difficult to get the microorganisms completely incorporated into the dirt without actually excavating the area and mixing it in manually. Also, the bacteria rely on oxygen to survive, and in some compacted, polluted soils there was simply not enough oxygen available for the organisms to live. Singer was surprised to learn that, in spite of the widespread knowledge of earthworms' ability to aerate the soil and transport microorganisms, few people had given any thought to using worms to incorporate specific bacteria into the soil for a clean-up project.

For his experiments, Singer chose an anecic worm—not *Lumbricus terrestris,* but *Pheretima hawayana,* a large fishing worm sometimes called a tropical nightcrawler that has also gone by the nickname California Golden and Alabama Jumper. This is a deep-burrowing worm that Singer knew would move constantly between the soil surface and the deep underground layers, bringing dirt up in the form of castings, where it could be repeatedly inoculated with the bacteria. He was also able to test the soil where the earthworms had been introduced to show that they were increasing the oxygen level belowground. In each of his experiments, more PCBs broke down when earthworms were present. The degradation of PCBs was also much more consistent throughout the entire soil area thanks to the earthworms. Finally—and this is perhaps the most extraordinary outcome of his research—he found that PCBs broke down faster when earthworms were present, even if no PCB-degrading bacteria had been introduced into the soil. He assumed that this was due to the fact that earthworms create a soil environment that is naturally more rich in microorganisms of all kinds, and that those microorganisms might have helped to break down the pollutant.

Singer used a fairly high concentration of earthworms for his experiments. It's not always so easy to get a worm population established in the wild, which is one of the major challenges scientists face when they're trying to use earthworms to carry out certain jobs. Right now, PCB-contaminated soil is most often excavated when it is discovered, and stored in large containers. Eventually, they would be taken to a landfill or burned. These soil-filled drums, Singer pointed out, are perfect for introducing large populations of earthworms. The environment can be controlled so that the conditions are just right for the earthworms to thrive, and scientists can expect fairly consistent results from each container.

There was one more conclusion to Singer's research, not much more than a footnote, but it points out the new directions in which earthworm science may be headed. While he was measuring the soil to see how much the earthworms increased oxygen levels belowground, he made an unexpected discovery. Soil that had a large earthworm population seemed to break down methane at a higher rate than soil without earthworms. Because methane is one of the most common greenhouse gases, it is widely believed to be a major contributor to global climate change. It is no secret that organisms in the soil can transform methane: peat bogs have a unique ability to consume methane and carbon dioxide from the atmosphere. They also give off methane, and some scientists speculate that there is a fragile balance between the organisms that emit methane and those that consume it in bogs. In the same way, earthworms help create soil conditions in which this balance exists. All Singer can say at the moment is that the notion of earthworms helping to reduce greenhouse gas is worth looking into. When he published his findings, he wrote, "The proposed effect of earthworms on enhanced methane consumption has relevance to global environmental change and is the subject of further study in our laboratory."

IF ANDREW SINGER is cautious about the potential for earthworms to clean up pollution and reduce greenhouse gases, he has every right to be. Earthworms are living creatures, fragile and temperamental, and experiments using them do not always turn out the way researchers expect. Over the last few decades, scientists have looked into the possibility of using earthworms to reclaim abandoned strip mines and rock quarries, where a chunk of earth, sometimes half a mountain, has been hauled away, leaving a hole that looks like an open wound. The idea was that earthworms could come in and close the wound, stabilizing the soil, working leaf litter into the ground, and making it possible for plants to get established. Depending on the conditions at the site—often the topsoil has been removed, exposing a layer of subsoil that simply won't support plant life— it can take years for a substantial earthworm population to build up on its own, but by using the techniques already made popular by organic farmers—minimal tillage, cover crops, the addition of compost and manure—worm populations could build up quickly and, in turn, help bring about natural restoration of the land.

Bringing worms into these kinds of damaged sites presents the same problem that farmers face if they want to introduce worms to their soil: it is not easy to find the right worm in sufficient quantities. Also, the soil might be too acidic for most species of earthworms. Finally, introducing earthworms into a climate that is either too hot or too cold will be self-defeating. It's a slow process. Even when earthworms have been brought into abandoned strip mines and rock quarries, it has taken up to ten years for them to fully establish themselves.

I talked to Jack Vimmerstedt, an expert in strip-mine reclamation at Ohio State University. He published several papers over the last few decades about the potential for using earthworms in these damaged areas. Ohio coal-mining laws required land to be restored after the mining was complete, and Vimmerstedt speculated that earthworms such as *Lumbricus terrestris* could be used to help build up sugar maple forests after strip mining. "I figured they'd bury the leaf litter, bring new soil up to the surface, and just help to build up the earth," he told me. "And they did. Sugar maple, European alder . . . those leaves are like ice cream and candy to worms. They ate it up. Except . . ."

I was already thinking about the exotic worms in Minnesota's forests. "Except the forests in Ohio probably didn't have night-crawlers to begin with, right?"

"Well, that's right," he said. "*Lumbricus terrestris* has gotten a bad name for itself when it comes to forests. It's been tagged as an invasive species. So maybe it's not such a good idea to fill a forest with nightcrawlers. Besides, now the state would rather see strip mines turned into pasture instead of forest."

"Worms are supposed to do great in pasture, aren't they?" I asked.

"Sure," he said. "We tried introducing worms to help get new pastures established, but when we went back two years later to check on them, we couldn't find any. Why? Well, one reason might have to do with the fact that they stockpile the topsoil and bring it back to the site after the strip mining's done. Some of the spoil material is rich in pyrite, and it's just too acidic for the worms. That could be one of the problems. Besides, it's just hard to tell what worms might do in a new environment."

Eventually, he reached the conclusion that there might not be a role for worms to play in the reclamation of strip mines. But that's only because he was looking at the short term, thinking in increments of ten years instead of a million years. In the long run, earthworms can reclaim strip mines, along with everything else. They will continue their slow, incremental work, consuming earth, bringing castings to the surface, and creating, over time, a new landscape.

SOMETIMES I WONDER if it is too much of an imposition on earthworms to push them into polluted ground, or to force-feed them a particular bacteria because we'd like to see it spread around. Darwin noticed that humans tend to exploit any characteristic for their own good, writing that "in the process of selection man almost invariably wishes to go to an extreme point." Are we taking advantage of earthworms? Shouldn't we clean up our own messes, or learn not to make them in the first place?

Earthworms are the custodians of the planet. They were here for millions of years before we came along. They survived the extinction that killed off the dinosaurs; I imagine they'd do just fine if something came along and wiped us out, too. *Eisenia fetida* may have grown particularly accustomed to food supplied by humans, but most of the species of worms around the world have little contact with us.

Darwin realized that earthworms, collectively, were a force to be reckoned with. Whether or not it is ethical or wise for us to enlist their help in fertilizing our farms, or cleaning up our pollution and garbage, we should remember one thing: we need worms more than they need us.

# The Ascent of the Worm

Will the stars from heaven descend?
Can the earth-worm soar and rise?

—A. L. Gordon, *Ashtaroth: A Dramatic Lyric,* 1867

I CAN'T DECIDE if this is the ultimate insult to an earth-worm or its highest calling, but it has been suggested that worms may be perfectly suited to play an even more intimate role in the lives of humans—that is, to help process sewage.

Bruce Eastman, a manager at the Orange County Environmental Protection Division in Florida, began working on a way to use earthworms to treat human sewage. He knew that worms, through their digestion, killed some microbes and encouraged others to flourish. It seemed plausible that worms could work their way through sewage, digesting harmful bacteria and shifting the microorganism population around so that the end product was safe for plants and even food crops. When I first heard about the idea, I had the same reaction that most people probably would have: I was repulsed by the idea of using human waste in my garden, even after earthworms had digested it. But then I thought about the bags of manure that I've hauled home from the nursery over the years. What's the difference? If earthworms could make sewage safe, if they could turn it into vermicompost, why not use it in the garden?

When Eastman began his research in 1997, only a handful of sewage treatment plants in Florida were able to produce what the EPA calls class A biosolids, or sewage sludge that is free of harmful bacteria. Most of the sewage produced in Florida—about two hundred and thirty thousand metric tons at that time—was spread on unused land for want of a better way to get rid of it. Over time, the rules for disposal of this waste have gotten stricter, and sewage treatment plants were facing the possibility of a very expensive retrofit to meet the new requirements. Eastman started to wonder if earthworms could reduce the dangerous pathogens in biosolids—namely salmonella, *E. coli,* and other gastrointestinal viruses. Experimenting on a test row, he and his colleagues found that the bacteria started to disappear when earthworms were given a chance to eat their way through the sludge. The EPA was interested in the idea and granted Orange County an experimental permit to do a larger study. The goal was to see if earthworms could bring about a three- to fourfold reduction in pathogens, which would be enough to ensure that the sludge was safe for human contact.

The researchers set up long windrows—gently mounded rows —of biosolids at the sewage treatment plant. They measured the levels of pathogens in each row to make sure they were comparable, then added the red wiggler, *Eisenia fetida,* to half the rows. In just six days, they had their answer: earthworms far exceeded the targets set by the EPA, bringing harmful bacteria down to a level that would earn the biosolids the cleanest rating, class A. Class A biosolids can be safely applied to land for all kinds of agricultural uses (one of the EPA's booklets includes a photograph of a flowerbed at Walt Disney World's Epcot Center that has been enriched with biosolids). Since then, similar results have been achieved in sewage treatment plant studies around the world; in fact, one plant in Korea processes eighty tons of sludge every day in a giant earthworm reactor. Although the EPA has not yet approved a method for earthworm composting of biosolids, the use of some kind of biological process to treat sewage has become increasingly popular, and it seems likely that earthworms may have a role to play in the near future.

You may never tour a sewage treatment plant yourself; I can't blame you for that. But when I heard that a worm composting project for biosolids was just getting underway at a new treatment plant in Pacifica, a town just south of San Francisco, I had to see it for myself. My uncle, David Sands, was in charge of a native plant restoration project at the facility, and he'd been asked to head up the worm composting project. One afternoon I put on my oldest clothes—a pair of jeans and a flannel shirt that I would not mind throwing away at the end of the day, if I felt so inclined—and I drove down the coast to take a look.

The Calera Creek Water Recycling Plant is situated off Highway 1 just south of San Francisco. The turnoff is unmarked; it could be any one of a dozen roads leading off the highway, through fields of artichokes or Brussels sprouts, to one of those unspoiled beaches known only to the locals. But when I arrived, I realized that this road was too newly paved to be a local surf hangout. There was a recently striped parking lot to the left with the correct number of disabled-accessible spaces—clearly a city lot. I drove through the gate and up a slight hill and there it was: wilderness. An expanse of chaparral bushes and dry grass. A creek, overgrown with willows, winding to the ocean. Here was the city's state-of-the-art sewer treatment plant, designed just the way the neighbors wanted it: underground, where it could be neither seen nor smelled. I drove down the hill, where only the employee parking lot and the front offices were visible. Everything else was cut into the hillside or buried underground entirely.

That's not all that distinguishes the Calera Creek Water Recycling plant from its peers, I found out. No chemicals are used in the processing of the waste: naturally occurring bacteria and ultraviolet light do all the work. The plant's capacity is flexible. Often it can power down during the day and power up at night, using more energy during off-peak hours, a real boon in California's ongoing energy crisis. Aboveground, recently planted native plants stabilize the hillside. Clean water flows out of the facility into a restored wetland that is home to great blue herons, San Francisco garter snakes, and California red-legged frogs. Bike and pedestrian paths run to the ocean, where water from the plant is discharged after receiving a final polishing in the wetland. The solids (sludge) emerge from the facility as pathogen-free, class A biosolids that will, if the worm composting project proves successful, be digested by thousands of earthworms and turned into fertilizer. It's a far cry from the old plant, where hazardous spills into the ocean and numerous safety violations led the state to shut it down.

I wasn't prepared for how clean and modern the plant would be. I can't say for certain what image the phrase "sewage treatment plant" conjured up in my mind, but I know I expected a filthy place, damp and foul smelling. I had never stopped to wonder how exactly sewage got cleaned. It was just something I took for granted. Imagine: before the middle of the twentieth century, cities deposited their sewage—untreated except for a pass through a metal grate and possibly a gravel filter—into rivers, lakes, and oceans. And that was progress. When Darwin arrived in London after his voyage on the *Beagle,* the first central sewers, which would carry sewage away from homes and into waterways, were being installed. Before that, Londoners deposited their waste in cesspits under their homes. The stench was often overpowering, and the levels of bacteria were certainly responsible for the kinds of lingering illnesses from which Darwin suffered. It is hard to believe that such conditions prevailed only 150 years ago. And now, after so much progress, it seems a little ironic that the most modern sewage treatment plants are being built underground, as if we are once again burying our sewage in the domain of earthworms.

When I met Dave Gromm, the plant's superintendent, I asked him about its name. He laughed and shrugged his shoulders. "Calera Creek Water Recycling Plant? I guess somebody thought it would sound better than 'sewage treatment plant.' It's the same thing, though. It's still sewage. It's just what we do with it that's different." He'd set aside the afternoon to walk me through the plant and explain its design, which is innovative and unusual: although the various components are in use around the country, no one had ever put them together in quite this way.

Sewage, which includes household waste, some industrial waste, and runoff from storm drains all over town, flows to a pump station where large filters screen out inorganic objects—soda bottles, tennis balls, and anything else that made it down a sink or through the grate of a storm drain—and the pumps move the sewage to the plant, where it fills one of five underground chambers called "sequencing batch reactors," or SBRs. Putting these SBR tanks underground is what made it possible for the plant to be so well hidden from view. Gromm was careful when he mentioned this key element in the design, which was included to please neighbors living on either side of the plant. "From an operator's standpoint? It makes the tanks hard to get to. Fiberglass hatches would've been nice." I couldn't argue with him. If I was in charge of five tanks filled with 1.2 million gallons of raw sewage each, I'd want to be able to see them, too.

On the day of my visit, a quiet Friday afternoon, Gromm took me downstairs to watch the process from a computer terminal. I looked over his shoulder. "Look, this one's filling now," he said. "The bacteria have been sitting in the bottom of the tank for a while now, and they're hungry." The image on the screen was like a sewage treatment plant operator's version of a flight simulator. Each part of the system was illustrated by animated schematics: green dots indicated that a chamber was open; a grey mass at the bottom of the tank represented the bacteria.

Once the tank filled, we watched air being pumped in (on the screen, it looked like tiny animated champagne bubbles being pushed into the tank). This creates a reaction with the bacteria that converts ammonia to nitrate. After the air is pumped in, a vigorous mixing turns the nitrate into nitrogen gas, which is then released into a massive air filtration system that prevents any odor from leaving the plant.

The mixing stopped so that the solids could settle to the bottom of the tank. "OK, the bacteria are done eating. This is its quiescent period," Gromm said, with something that sounded almost like affection. The bacteria did their job in only seventy-five minutes and now the tank sat perfectly still so that the solids could separate from the liquids. "After this, it'll be ready for the decant. There's this device that sits just a little below the surface and draws water out. That way it avoids any solids that are still floating on the surface." From here, the water will be pumped to an ultraviolet-light disinfectant system and the solids are pulled into a holding tank. The whole SBR process takes about five hours, with five tanks in different stages at any given time.

Gromm left his computer terminal and led me through the rest of the plant, most of which is also underground but situated in large rooms where the equipment is accessible. The first room, called "the gallery," is adjacent to the buried tanks; this is where the sewage is pumped in and pumped back out, and this is also where the massive air filters remove any odor from the air leaving the plant. The filters do such a good job that people working in the front office can't detect any smell at all.

It stank in the gallery, but it wasn't as bad as I thought it would be. It was not as foul as the dump on a hot day; it did not even stink as much as the house that my college landlady shared with several dozen cats. The smell inside the gallery was an unpleasant, vaguely organic smell, just a few degrees off from that of a dairy farm. It was just powerful enough to make me glad that all the valves, seals, and pumps were in perfect working order. You wouldn't want to be here on a day like the one that Gromm described, shouting over the roar of fans and motors.

"This whole thing's computer controlled," he said, waving his arm at the equipment all around us. "We had all these glitches when we first got started. The blowers would stay on when the computer said they'd been turned off. Or the filters would clog. The thing is, we've got continuous flow here. You can't just take a filter off line to unclog it.

"One day early on, the computer gave the command to open the influent valve and release the sewage. But what the computer didn't know was that this key that opens and closes the valves had dropped out of place. It told the system to release the sewage but this valve was still closed and we had sewage"— he paused and rolled his eyes—"*everywhere*. I don't know how many times we flooded this parking lot."

"How'd you ever figure out what caused it?" I asked. We were looking at a fifth of the plant's computerized equipment, and it filled a room larger than a high school gymnasium. I could see dozens of places where a valve could get stuck or a filter clogged.

He beamed. "I watched it being built. They needed inspectors on the job site so I brought some of my guys down from the old plant and we did the inspections. There's not a pipe anywhere that I didn't watch being laid."

He took me to a large room upstairs where the solids get sent after they leave the SBR. "It gets deposited up here," he said, pointing to a tank of liquid sludge. "Then it gets sent over to the ATADs."

ATAD stands for "autothermal thermophilic aerobic digestion." It's a process that was developed in Germany and has been used in Europe since the 1970s but has only recently gained popularity in the United States. The EPA classifies sludge coming out of ATADs as class A, thanks to the thermophilic, or heat-loving, bacteria that kill the fecal coliform, the salmonella, and the roundworm eggs by heating it up to 160 degrees.

The smell was worse up here. I was getting a strong urge to wash my hands. Now I understood why the women's bathroom had two shower stalls and four kinds of disinfectant soap. The facility was clean, but just knowing what was getting pumped through here at the rate of a few million gallons per day could make anyone a compulsive hand washer.

The biosolids spend nine days in the ATADs, working their way from one digester into the next, before they are pumped into a centrifuge machine that extracts more water. The solids leave the centrifuge by means of a chute and are deposited into trucks waiting in a truck bay downstairs.

Gromm took me there next. A couple of guys were using large rakes and shovels to smooth out the biosolids in the open beds of their trucks. "What you're smelling is ammonia," he said. "That last little bit of water that comes out of the centrifuge? That's incredibly high in ammonia. We just feed it back into the system at the starting point, so it goes into the SBRs along with the rest of the sewage. I don't think we'd counted on all that extra ammonia at first. There's more ammonia throughout the whole process because of that little bit we add back in. We're thinking about adding a step where we cool the biosolids back down and let the nitrifiers go to work on them to get the ammonia down, then put it through the centrifuge. The bacteria that eat the ammonia just can't survive at these higher temperatures."

Clearly, this is a biological process that requires constant adjusting. Gromm manages a colony of bacteria along with a crew of workers. The guys with the shovels didn't look too unhappy, considering what they were shoveling. "You know what this stuff is?" Gromm asked me, and I nodded. "Well," he said matter-of-factly, "it looks like what it is." The chute opened every few minutes and deposited another black, stinking load into the trucks.

This was the end of the line for the biosolids. It was now a good, clean, class A product that would get trucked to a landfill. The city would rather find a use for the solids and eliminate the nearly $100,000 per year it spends in disposal fees. That's where the worms come in. But before I went to take a look at the worm composting project, Gromm wanted to show me what happens to the water once it separates from the solids. It's a good example of the kind of "green" solution that's possible at a treatment plant like this one.

We walked out of the vehicle bay into the parking lot. Across the lot, a fence surrounded long rows of concrete walls, sunk down into the ground so only the tops were visible. Water poured over each wall. The overall effect was not unlike a water feature in a city park. This, Gromm explained, is the sand filter, where larger particles are removed from the water before it flows to the ultraviolet-light treatment belowground.

Over three million gallons of water pass under the closely spaced ultraviolet lights each day. This process allows coliform to be removed without the use of chlorine. Flow regulators monitor the rate at which water is discharged into the network of irrigation pipes leaving the plant. "You know, there's good coliform, too," Gromm said with what I was coming to see as a healthy respect for bacteria. "It's not all harmful. Some of it helps rivers and streams to survive and hosts all kinds of life. But we take it all out anyway. Sometimes I wish we could only take out the bad stuff and leave everything else."

Gromm and I made one more stop back inside the plant: a sparkling clean, sunny laboratory—one of the few aboveground rooms—where the water and biosolids are tested for compliance with EPA standards. Beakers on the counter contained water from every step of the process. Gromm held up a beaker of absolutely clear water. "This is the final product," he said. "The EPA requires that we have ten milligrams per liter or less of suspended solids and BOD (biochemical oxygen demand). Last week, we had one milligram per liter. A couple days ago it was zero. Same thing with ammonia. We're under one milligram per liter steadily."

Gromm manages a delicate biological system, one that can be upset by somebody dumping oil or gasoline down a sink, or by a sudden storm pushing excess water into the pipes. The process is extraordinarily complex, but in some ways it is as simple and as natural as what an earthworm does when it works through a compost pile. Without the aid of chemicals, a community of microscopic creatures digests, reproduces, and transforms an organic substance—human waste—into something quite different. Nothing but biosolids and clean water leaves the plant.

. . .

THE WATER FLOWS into what most people would agree is the crown jewel of the Calera Creek Water Recycling Plant: a man-made designer wetland that is home to endangered species of birds and amphibians, and a hike and bike trail that draws people to the plant. Building this wetland was more than an enormous landscaping project; it was an attempt to enlist bog plants, snakes, frogs, birds, and insects in the final cleaning— "polishing"—of the water before it flowed into the ocean. It seemed like a perfect complement to the biological processes at work inside the plant. And just as I had come to realize that there was more going on in the soil under my feet, I now saw that more was at work in those low-lying, swampy stretches of coast than I had ever realized. Wetlands do something—they breathe, they clean, and they transform.

The idea of releasing treated wastewater into a wetland is not unique to Pacifica. Several dozen natural or constructed wetlands are in use at sewage treatment plants around the country. Wetlands have proven to be the ideal environment for the discharge of treated wastewater: they act as a sponge to absorb excess runoff, and the marsh plants, microorganisms, and silty soil all filter and polish the water. In fact, wetlands can perform, to some extent, all of the functions of a conventional wastewater treatment plant. By the mid-nineties, the EPA had identified wetlands that do just that, such as the Congaree Bottomland Hardwood Swamp in South Carolina, which carries out all the functions of a $5-million tertiary treatment plant. Now a constructed wetland has become a citizen committee's dream solution to the problem of discharging treated wastewater, or "recycled water." It is an acceptable place to send the water—

preferable, say, to watering a school playground with it or pumping it through a fountain in the town square, practices that would be entirely safe but unpalatable to most people. With wetlands in the United States vanishing at an alarming rate—about sixty thousand acres per year—cities are all too willing to construct a wild and nearly self-maintaining green space that can receive the discharged water, host endangered wildlife, and accommodate a few bicyclists and bird watchers. Constructing a wetland was a natural choice for Pacifica, and the site, an abandoned rock quarry where Calera Creek once flowed, was begging for a restoration of some kind anyway.

PACIFICA HAD NEVER attempted anything on this scale before. "I had my sewer crew growing native plants for the wetland," Gromm said sheepishly. "What do they know from plants? That's when one of my guys went out and found your uncle."

My uncle David makes a living growing plants that are native to a very specific stretch of coastline just south of San Francisco. "You can't grow these plants in Fresno and ship them here," he told me as I took leave of Gromm and joined him for a walk through the wetland area. "You have to grow them right *here*," and he pointed to the ground. "You see all those plastic pots above where the SBRs are buried? That's my nursery. That's where we grew the plants for the wetland."

I walked the site for the first time with my uncle in 1998, just after he'd started the job. Public works crews had cleared all the vegetation off the site—about one hundred and ten thousand cubic yards—and excavated the area according to a detailed wetland design plan. Calera Creek had been realigned to flow along its original path. David and his crew had planted over fifty thousand native plants.

Frankly, it didn't look like much back then. At that time, the treatment plant itself was still under construction. Trucks rumbled up and down the hill and dust was everywhere. The native plants still looked pretty scraggly and only a few wildflowers had emerged from the mixture of straw mulch and seed that had been sprayed on the hillsides. The creek bed had been carefully sculpted to look natural, widening into ponds in a few places and narrowing down to a trickle in others. Bare sticks emerged from the water every few feet; my uncle explained that these were willow branches that would take root over the winter. I was skeptical. They looked like dead sticks to me.

"There's more going on here than you think," he insisted. A group of wetlands scientists had used a new system for assessing wetlands that was designed to categorize the wetland by its functions, identify other sites that functioned the same way, and create ways to monitor what was happening over time. "People used to stick plants in the ground and call it a wetland," one of the consultants on the project told me. "Like it was a kind of glorified gardening. But there's more to it than that."

David pulled out a muddy blueprint of the thirty-acre area between the new treatment plant and the sea. The wetland was divided into discrete sections called polygons, with a specific plant list for each. The plant lists were developed after the consultants visited dozens of other wetlands along the coast and made detailed lists of the plants growing there. "Each polygon gets planted with a particular plant community," he said. We were standing in Polygon 41, Palustrine Forest II, Point Bar. "It's pretty technical, but it's got to be messy and uneven, too. I had to keep reminding the guys not to plant in straight rows. What we're trying to grow here is chaos. It's got to be wild."

Now, TWO YEARS LATER, I didn't recognize the place. I couldn't get anywhere near Polygon 41. Standing on top of the buried SBR tanks, I looked down on the tops of willow trees, their canopies so dense that I couldn't see the creek at all. Sprinklers placed high on the opposite hillside sprayed water over the expanse of green. "All that water comes from the treatment plant," my uncle told me. "Usually when you do a project like this you never have enough water to get the plants established. But that's not a problem here. They're talking about sending some of the water over to a golf course, but until they do, we get all the water we want." As if to prove his point, a set of automatic sprinklers came on next to us and started watering the hundreds of potted plants he had situated near the SBRs. "Those are for our next project," he said. "This one turned out so well, now we're restoring the next creek down the road. Flood control project."

The dense vegetation over the creek has helped shelter endangered species. "I haven't seen a garter snake there yet, but they're hard to see anyway," he told me. "Red-legged frogs are showing up, especially down in the creek where they're hidden from birds."

The monitoring plan calls for regular visits to the wetland over a five-year period to measure its success in the form of the water quality, survival of trees and shrubs, and the presence of wildlife. Once the five years have ended and the creek is entirely overgrown, the wetland will be left more or less on its own, with Gromm's treatment plant crew watching the discharge of water and bird lovers walking the trails with binoculars, counting birds.

WHAT DOES ANY of this have to do with worms? As I toured the plant and walked the hillside with my uncle, looking down at the wetland below us, I realized what had happened here. Nature had been reengineered, harnessed, hired to do a job. If you allow a creek to go back to being a creek—if you let the trees and the bramble get overgrown, and you let the stream overrun its banks whenever it wants to—the wetland will take care of itself. The water that trickles into the ocean will be clean and pristine, if everything is just left alone to work the way it was designed to work. Earthworms have shown that they can take care of the soil in the same way that a wetland takes care of the water. Nature regenerates, it cleans, it hides a multitude of sins.

Does that give people an unlimited license to pollute? No, certainly not. But it offers a new way to look at how we manage our own waste. That's not to say that the room on the back of my house that might have once served as an outhouse suddenly seems more appealing—I still prefer the gleaming porcelain and chrome of the bathroom upstairs—but I can understand now that the advancement of science, along with a more thorough understanding of the extraordinary powers of such natural processes as a worm moving through the soil, could suggest a more practical and effective way of managing waste.

"WATER RECYCLING PLANT" is a fairly accurate name for the Calera Creek plant considering that ninety-five percent of what the facility processes is water. But it's no easy feat figuring out what to do with the other five percent, the biosolids. Because the finished product is so clean, its use is almost completely unrestricted. It can be spread on farm or forest land, used to fertilize plants in city parks, or given away in bags to local gardeners. The EPA's literature is effusive on the subject of biosolids' possible uses: "Nutrients found in biosolids, such as nitrogen, phosphorus and potassium and trace elements such as calcium, copper, iron, magnesium, manganese, sulfur and zinc, are necessary for crop production and growth. The use of biosolids reduces the farmer's production costs and replenishes the organic matter that has been depleted over time. The organic matter improves soil structure by increasing the soil's ability to absorb and store moisture."

Biosolids may have plenty to recommend them, but the city of Pacifica, like many cities across the country, has not had an easy time giving the stuff away. For one thing, it stinks. "You know when that truck's headed to the landfill," my uncle said. "You can smell it coming and you can smell it going." The other problem is that people just don't want to spread human excrement in their flower beds or artichoke fields. Cow manure's one thing. This is something else again.

Pacifica carts its biosolids off to a landfill at a considerable expense. Just giving it away would be an improvement; selling it to gardeners for a few dollars a bag, or selling it by the truckload to farmers, could even be a moneymaker. After all, Pacifica is at the northern end of a stretch of rich agricultural countryside. Most of the roses grown in the United States come from Half Moon Bay, just down the road. Cut flowers, strawberries, lettuce, pumpkins, artichokes, and Brussels sprouts all flourish along the coast and could benefit from a cheap source of good fertilizer, if it could only be made palatable to them.

That's where the worms come in. If earthworms go to work on the biosolids after they leave the plant, the smell will be reduced, the texture will be more even, and the final product will be even more nutrient-rich. Above all, the concept is more attractive to the public. Language is everything here: earthworm castings from the water recycling plant sounds vastly more appealing than sanitized human waste from the sewage treatment plant. Farmers and gardeners might want it, and parks and schools might take some for their gardens.

David told me about this plan over lunch. "You know I say yes to whatever those guys ask me to do," he said of the public works staff at the city. "And I've never regretted it. They want me to build a wetland, I build a wetland. They want me to find a way to help the fish get downstream, I've got guys trucking fish three days a week. Now they want a worm farm, and that's what I'm going to give them. Is there any special kind of worm I need?"

I explained about composting worms like *Eisenia fetida* and how they're different from the nightcrawlers that he was used to seeing in the soil. We talked about how the biosolids should be arranged: most worm farms place their feed in windrows, long rows about four feet wide and two feet tall. The worms start at one end and as fresh feed is added at the other end of the row, they work their way towards it. I toured a farm in Washington once where they process cow manure like this. The rows are laid in a U-shape; once the worms are finished at one end of the U, the castings are put through a harvester that spins it around and shakes out any remaining worms. That last step isn't even necessary, I told him, if the city didn't mind a few stragglers in its final product.

"You can buy a reactor and set it up in a warehouse if the city's got the money," I told him. "But if this is just a pilot project, there's nothing wrong with worms in windrows."

We went back to the treatment plant, where he'd picked an unused area to dump a load of biosolids. Finding a good location over the long term would be the biggest challenge. Each time he added fresh biosolids, the stink from the ammonia could be smelled a few miles away. It only lasts a day or two, but that is long enough to upset the neighbors. If they ever wanted to compost all the biosolids produced by the plant—several tons a week—a more secluded location would have to be found.

I knew from my talk with Bruce Eastman in Florida that the main challenge facing the city of Pacifica would be to make sure that the worms had a good food source. Class A biosolids that leave the plant at 160 degrees are basically sterile. The bacteria in the biosolids, both good and bad, have been killed, and these worms rely on living organisms such as bacteria as a major part of their diet. Nobody's entirely sure if class A biosolids offer a good enough food source for the worms. Eastman suggested purchasing an inoculant from a laboratory but wondered if reintroducing bacteria would somehow change the status of the biosolids from class A to something lower. I called a waste management consultant who suggested mixing the biosolids with grass clippings or food waste to make sure the worms had something familiar to eat. David decided to try mixing green waste from one of his landscaping projects into the biosolids, which would decompose and offer the worms a food source.

That wasn't the only issue. Worms are pretty picky about their environment, too. They prefer a temperature between sixty and seventy degrees. David would need to let the biosolids cool down before he introduced the worms to them. He'd also need to bring his pH meter and test the biosolids to make sure they weren't too acidic. They like it damp; he'd need to water the pile during dry months. Finally, though, I knew that worms were sensitive to salt and ammonia content, and that's where the real problem was. Last time the biosolids were tested, the ammonia was too high, and they had no idea what the salt content might be.

It can take a few weeks to do a soil test and get the results back, but David was eager to get started. Before I arrived, he'd dumped a load of biosolids on the piece of land he'd set aside, watered it, mixed in grass clippings, and driven a couple hours to Davis to pick up fifty thousand worms. "We can't ship in this heat," the worm farmer had told him, but David was all too happy to drive out to pick up the worms himself and inspect the farmer's operation while he was there. He returned with three wax-lined boxes, each punched with tiny airholes and taped securely shut. Inside, tiny red worms—mostly juveniles, without the fully developed clitellum that marks an adult—squirmed together in a tight ball of worm bodies. They were packed in coir (shredded coconut fiber), but the boxes held far more worms than coir. They were heavy with worms.

The worms sat in the back of the truck while David and I walked over to check out the pile of biosolids. He handed me a pair of rubber gloves and suggested that we test the temperature before we added the worms. I hesitated for a minute, realizing that I had just been asked to put my hand into a pile of Pacifica's collected excrement. But then I decided that I wouldn't ask the worms to do anything I wouldn't do myself. I put the gloves on and plunged my hand in the pile. It was hot to the touch. There was no way the worms would tolerate it. They'd have to spend the weekend in David's garage.

The next week, the pile had cooled and David buried the worms in it. They were still alive and moving through it a month later. He'd set up a second pile, without worms, for test purposes: the city would run soil tests once a week to see if the ammonia naturally dissipated along with the smell. If so, the ammonia problem might just be solved by letting the pile sit for a few weeks first. The windrow design would be ideal for this: the worms would simply not enter the most recently added biosolids until they had cooled and until the ammonia level had dropped.

THE STAFF AT THE plant continues to work on the ammonia levels, and David is monitoring the worms as they work their way through the rows of biosolids. He tells me that he's lucky to have enough time, enough land, and all the biosolids he can use. He can keep experimenting with each new load until he figures out what works for the worms. This is unfettered research, a process of trial-and-error and discovery, and in that way it reminds me a little of Darwin's work, of his tireless experimentation with worms in jars.

People at the plant seem optimistic about the project. The food source is free. After the initial investment in earthworms, they will reproduce quickly to meet the increasing supply, and the finished product will surely be a fine soil amendment. There are still a few challenges ahead—if worms are going to consume all the biosolids the plant produces, they'll have to find another location, one that isn't upwind of the town's residents. And for treatment plants that don't produce class A biosolids, it remains to be seen whether the EPA will approve a process that uses worm composting to reduce the harmful pathogens.

I should also point out that people still have plenty of concerns about growing food in biosolids—in particular, the possibility that heavy metals can accumulate in it. The solution seems to be to prevent factories (and people in their own homes, for that matter) from pouring toxic substances down the drain in the first place. Every cleaning product, every paintbrush that gets washed out in the sink, every spill of oil or gasoline that runs down the gutter, eventually ends up at a sewage treatment plant like Pacifica's. The level of heavy metals in sewage has declined over the last few decades, as industries are forced to find other ways to deal with their waste and the public is educated about the safe disposal of paints, solvents, and chemicals. A good biosolids composting project will have to monitor the level of heavy metals in the finished product. Earthworms can even play a role here as a biomonitor, allowing scientists to watch for long-term accumulation of heavy metals where biosolids have been used.

So there is more work to be done on all fronts. Still, it seems fitting that worms could find work turning a town's waste back into something that local farmers and gardeners can use. In doing so, the worms exercise their transformative power. They are near the bottom of the food chain, a meal for fish and birds, while humans eat from the top of the food chain, consuming an astonishing array of what lives on the planet. But eventually, even we become food for worms. Shakespeare saw this connection, writing in *Hamlet,* "A man may fish with the worm that hath eat of a king, and eat of the fish that hath fed of that worm." Should it come as any surprise, then, that earthworms have the power to transform human waste back into soil, where the cycle starts over again?

ALL THE TIME I spent at the sewage treatment plant got me thinking about something my husband, Scott, told me recently. He wants to build a chicken coop and get a couple of hens. He likes the idea of fresh eggs, and now I am attracted to the notion of all that chicken manure.

But soon I realize that the very best soil in the garden will be right under the chickens' enclosure — the one place where I couldn't grow anything. Imagine, all that chicken manure landing on the ground, and all those earthworms rising to the surface to eat it. The soil under those chickens will be the best soil I've ever seen. Shoveling out the manure isn't enough. I need to find a way to make use of that hen-scratched, manure-laden earth under the coop.

I do a little research, and pretty soon I find plans for a portable chicken coop called a chicken tractor that can be moved every spring to a new spot in the garden. I walk my vegetable garden, mapping out a new design that will allow the vegetable beds to be grouped together into four ten-by-ten beds. One of those beds will be just the right size for a small chicken coop. Every year, the chickens will move to a new bed, and every year, I will come along behind them and plant vegetables in that manure-and-earthworm-rich mixture. The chickens themselves are interesting to me, but they will be Scott's pets. I am more excited about what this means for me and the worms.

When it comes right down to it, my worms aren't heroic or extraordinary in any way. They won't solve the world's pollution problems or treat sewage (apart from the chicken manure) or eat anyone's garbage other than my own. They'll just stay here in this patch of earth, along with me, and try to make the best of the environment they live in. Generations of worms will live on in the soil, long after I'm gone, long after this old house has fallen down. But they will renew the earth. What could be more extraordinary than that?

I GO BACK TO DARWIN once more, at the end of his life. In his final years, he seemed to look forward to joining the worms underground. Around the time *The Formation of Vegetable Mould* was published, he wrote to a friend, "I have not the heart or strength at my age to begin any investigation lasting years, which is the only thing which I enjoy; and I have no little jobs which I can do. So I must look forward to Down graveyard as the sweetest place on earth." He was perhaps better acquainted than anyone else with the fate that awaited him, with the life cycle of which he would, in death, play his part. This did not seem to trouble him. He wrote that he had "no fear of death, after such a life." One biographer wrote that when his hired gardener turned the compost pile, exposing the burrows of earthworms, Darwin "momentarily glimpsed his grave."

He died on a spring afternoon in 1882. When it was announced that he would be laid to rest in the family plot near his home, a public cry went out to bury him instead in Westminster Abbey. One newspaper acknowledged that "Darwin died, as he had lived, in the quiet retirement of the country home which he loved; and the sylvan scenes amidst which he found the simple plants and animals that enabled him to solve the great enigma of the Origin of Species may seem, perhaps, to many of his friends the fittest surroundings for his last resting place." Still, the newspaper argued, his proper place was not at the Down graveyard, in the company of his beloved earthworms, but in the Abbey, "among the illustrious dead." There was strong popular support for this idea, and about a week after his death, Darwin was buried next to astronomer Sir John Herschel and near Sir Isaac Newton. There was little debate among the church leadership over the appropriateness of this decision. In a sermon following his burial, the Bishop of Carlisle described as foolish the notion that "there is a necessary conflict between a knowledge of Nature and a belief in God."

In spite of this small reconciliation between Darwin and the church, it was no secret that Darwin's faith during his lifetime was shaky at best. His work necessarily challenged some of the most deeply held teachings of the church, and he often struggled to keep his ideas private because he knew how inflammatory they would be. Notwithstanding the bishop's kind words, the fact is that the conflict between a knowledge of nature and belief in God threatened the reputation of his family and even shook the foundation of his marriage to Emma, who hoped to spend eternity with her husband but feared she would not.

If Darwin had any notion of heaven, he surely believed that it was all around him. Adam Phillips in *Darwin's Worms* suggested that in his study of earthworms, Darwin found for himself a kind of immortality, a kind of redemption, and a certain sly delight in the notion that worms created the earth. Phillips wrote that earthworms "preserve the past, and create the conditions for future growth. No deity is required for these reassuring continuities. . . . Darwin has replaced a creation myth with a secular maintenance myth. This is how the earth maintains itself, as fertile and ongoing." Perhaps Darwin realized that the promise of eternal life, of resurrection, had been delivered. It was already happening, right beneath his feet.

# EPILOGUE

PEOPLE OFTEN ASK ME about getting started with worm composting. After spending seven years as a keeper of worms, and a solid year or two engaged in a study of their habits and history, I still say that there is not a finer pet anywhere. I've had dogs that would not walk on a leash and birds that refused to sing, and at present I've got two cats who nap when they should be chasing mice and chase each other around the bedroom at night when I'm at risk of nodding off to sleep myself.

But a herd of worms will earn its keep, no doubt about it. They'll take care of the garbage, fertilize the lawn, and bait your fishhook if you like to fish. They'll provide the kids with science-fair projects and show-and-tell offerings for years to come. And they'll do it all with a minimum of fuss and expense.

Plenty of books give advice on worm composting, and I won't try to summarize everything they have to say here. Instead, I'll tell you what I would want to know if I were setting up my first worm bin right now, and you'll learn the rest as you go.

The first step is choosing a bin. It's not difficult to locate instructions for making a worm bin out of scrap lumber or plastic storage tubs. After a trip to the hardware store and a stop at the bait stand, a homemade worm bin can be yours for twenty bucks, maybe forty.

But those homemade bins have a few downsides. Drainage is

one concern. The worms won't sit for long in soggy food, and you won't like the smell if they do. It's often hard to get a home-made bin to drain properly from the bottom and ventilate prop-erly from the top, all while keeping the fruit flies out and the worms in. Harvesting castings is another problem: if the worms live in a single bed together with their food and castings, it can be ridiculously difficult to extract them. Some people say that if you shift the food to one side of the bin, the worms will follow, leaving the other side more or less vacant. But worms do tend to wander, especially within the confines of a small space, and it can take weeks for them to abandon one area of a small bin en-tirely, if they ever do.

There are many good commercially made worm composters out there—the Can-O-Worms, the Wriggly Wranch, and the Worm Factory, to name a few—designed to give worms the best living conditions possible and make it easy to harvest and use their castings. Some city and county recycling departments sell these worm bins at low cost; for instance, the county of San Mateo, where my uncle David lives, sells the Wriggly Wranch to county residents for a discounted price that makes it possi-ble for just about anyone to get into worms. In other counties, a recycling group or a university extension program might hold a composting or recycling fair and sell bins at a reduced cost for that day only. The Resources list in the appendix gives a few more suggestions for finding a suitable bin.

Once you've selected a bin, the next task is getting some worms. You can buy your worms from a bait stand as I did, but you'll be better off ordering worms from a worm farm, which the leaflet that came with my bin instructed me to do. You'll save a little money and you'll have the assurance that you are getting the right worms. Besides, worm farmers are a friendly group, always willing to answer a few questions from their customers. If you live anywhere near a worm farm, you might even get a tour. (Be forewarned that some worm growers, fearing for their worms' health, won't ship live worms during hot summer months.)

It's important to find a good location for the bin and introduce the worms to their new home as soon as possible. Some people keep their worm bins inside—under the kitchen sink, for instance—but one look at my husband's face told me that the worms would live outside and that if I let even a few into the kitchen, I might be sleeping out there with them. My composters sit on the back porch, where they are sheltered from the rain and easily accessible to receive the day's kitchen scraps. A garage is also a good location. Although my worms have never left their bins, one worm farmer I know includes a nightlight with every worm bin he sells, the idea being that even a small light will discourage worms from exploring your garage at night.

Most composters come with a coir brick that, when soaked in water, expands into a fibrous substance like peat moss. This goes in the bottom of the bin as a bedding material for the worms. If you don't have coir, damp shredded newspaper or rice straw makes a good substitute. The worms will settle into their bedding, adjust to life in the worm bin, and get ready to eat.

Worm bins often come with instructions telling you not to feed the worms for the first few days, but if you've made it this far, you probably won't be able to resist throwing a banana peel or a lettuce leaf their way to see what happens. That's what I did, and I doubt it did any harm. This is the best way to start feeding worms anyway: a little at a time, giving them only as much as they can eat in a day or two. At first, a pound or two of food a week may be all that they can handle, but pretty soon they'll be eating all your kitchen scraps, especially if you finely chop them so they'll decay faster.

My worms have definite likes and dislikes, demonstrated by which foods are left to rot untouched in the bin, and which are covered in a mass of wrigglers within a few hours of being offered to them. They love banana skins, melon rinds, and lettuce leaves. They'll eat coffee grounds and stale bread, but they won't touch onions, oranges, or anything too acidic. They can't eat fats in any form (including salad dressing), or meat, or dairy products. They do, however, like crushed eggshells because they provide a source of grit for their gizzards and help to moderate the pH level in the bin. No matter what you feed them, be sure to pile plenty of shredded newspaper on top to hold moisture in and keep fruit flies out. (Over the years, it has become a comforting part of my morning routine to deposit my coffee grounds and banana skins into the composter, then stand over them and tear up a section of the morning paper for them.)

Be prepared for a few critters to find their way into your bin no matter what you do. After all, the worms eat the microorganisms—bacteria—that occur in the bin as food scraps decompose. A few creatures that are large enough to be seen may also start to show up, and they are harmless additions to the system. Pill bugs—roly-poly bugs—are always attracted to compost piles and are considered a sign of good health in any composting system. Pot worms, or terrestrial enchytraeids, are often mistaken for baby earthworms. (Pot worms are white, but baby worms are reddish and faintly translucent.) The pot worms are closely related to earthworms and feed alongside them, causing no harm. And a small population of fruit flies may be inevitable, although they are also harmless.

Depending on the type of bin you have, your first harvest of worm castings could be weeks—or months—away. I rotate the trays of my Can-O-Worms about every three months, each time harvesting a few pounds of black, rich castings. By putting the tray to be harvested—the bottom tray, in my case—on top of the bin and leaving the lid off for an afternoon, the few worms that might remain in that lowest tray squirm away from the light and plunge down into one of the lower trays. I like to keep a stock of finished compost on hand, so my worm castings get mixed into a garbage can that also contains garden compost, manure, and other soil amendments from the nursery. I use this rich mixture—worm castings included—as a planting mix when I'm transplanting new plants, as an addition to potting soil, and as a layer of mulch in the vegetable garden. You can also dig worm castings directly into the soil or make a kind of compost tea by mixing the castings with water and using it to water your garden.

Most of all, though, if you keep worms, you shouldn't be afraid to get to know them. Forget any advice you may hear about leaving your worm bin undisturbed. While I was writing this book, I found I couldn't sit at my computer for more than an hour without going downstairs to check on the subjects of the book themselves. They are clean, quiet, well-behaved creatures, interesting to watch, and even beautiful in the way that any organism can be if you know just how to look at it. Put a worm in your hand and watch it expand each segment in turn, arch its back, flex its muscles. You will be won over. Lift up the top layer of food and newspaper in your bin and you will surely be awestruck at the spectacle of such industry taking place just outside your back door: thousands of worms churning through your apple cores and coffee grounds, your newspaper and dryer lint, taking it all in and turning it back into earth.

# SELECTED BIBLIOGRAPHY

Appelhof, Mary. *Worms Eat My Garbage*. Kalamazoo: Flower Press, 1997.

Browne, Janet. *Charles Darwin: The Power of Place*. New York: Knopf, 2002.

Conniff, Richard. *Spineless Wonders*. New York: Henry Holt and Company, Inc., 1996.

Darwin, Charles. *The Formation of Vegetable Mould, Through the Action of Worms, With Observations on Their Habits*. New York: D. Appleton and Company, 1897.

Darwin, Francis. *The Autobiography of Charles Darwin and Selected Letters*. Mineola, NY: Dover, 1958.

Dirks-Edmunds, Jane Claire. *Not Just Trees: The Legacy of a Douglas-fir Forest*. Pullman: Washington State University Press, 1999.

Edwards, Clive. *Biology of Earthworms*. London: Chapman & Hall, 1977.

———. *Biology and Ecology of Earthworms*. London: Chapman & Hall, 1996.

———. *Earthworm Ecology*. Boca Raton, FL: Lewis Publishers, 1998.

Friend, Hilderic. *The Story of the British Annelids (Oligochaela)*. London: Epworth Press, 1924.

Gaddie, Ronald, Sr. *Earthworms for Ecology & Profit*. Ontario, CA: Bookworm Publishing Company, 1975.

Hendrix, Paul. *Earthworm Ecology and Biogeography in North America*. Boca Raton, FL: Lewis Publishers, 1995.

Hubbell, Sue. *Waiting for Aphrodite*. New York: Houghton Mifflin, 2000.

Laverack, M. S. *The Physiology of Earthworms*. New York: Macmillan, 1963.

Lee, K. E. *Earthworms: Their Ecology and Relationships with Soils and Land Use*. New York: Harcourt Brace Jovanovich, 1985.

Minnich, Jerry. *The Earthworm Book: How to Raise and Use Earthworms for Your Farm and Garden*. Emmaus, PA: Rodale Press, 1977.

Morgan, John. *The 7th International Symposium on Earthworm Ecology*. Cardiff: Cardiff University, 2002.

Payne, Binet. *The Worm Cafe: Mid-Scale Vermicomposting of Lunchroom Wastes*. Kalamazoo, MI: Flower Press, 1999.

Phillips, Adam. *Darwin's Worms: On Life Stories and Death Stories*. New York: Basic Books, 2000.

Rodale, Robert. *The Challenge of Earthworm Research*. Emmaus, PA: The Soil and Health Foundation, 1961.

Samuelson, James. *Humble Creatures: The Earthworm and the Common Housefly, In Eight Letters*. London: John VanVoorst, 1858.

Satchell, J. E. *Earthworm Ecology: From Darwin to Vermiculture*. London: Chapman and Hall, 1983.

Smillie, Joe. *The Soul of the Soil: A Soil-Building Guide for Master Gardeners and Farmers*. White River Junction, VT: Chelsea Green, 1999.

Stephenson, J. *The Oligochaeta*. Oxford: Oxford University Press, 1930.

Voisin, Andre. *Better Grassland Sward*. London: Crosby Lockwood & Son Ltd., 1960.

Wilson, Eric. *Worm Farm Management: Practices, Principles, Procedures*. East Roseville, NSW, Australia: Kangaroo, 1999.

Wolfe, David. *Tales From the Underground: A Natural History of Subterranean Life*. Cambridge: Perseus, 2001.

# APPENDIX: WORM RESOURCES

## MAGAZINES AND NEWSLETTERS

### Worm Digest

A quarterly magazine published by a nonprofit in Eugene, OR. Its website features a lively worm discussion forum. Includes links to other websites of interest. Publishes *The Art of Small-Scale Vermicomposting*, a handy reference guide for home composters.

PO Box 544
Eugene, OR 97440
www.wormdigest.org

### BioCycle: Journal of Composting and Recycling

A professional journal on the subject of large-scale composting and recycling. The journal and related conferences sometimes feature information on worm composting projects.

The J. G. Press, Inc.
419 State Avenue
Emmaus, PA 18049
www.jgpress.com

### Casting Call

A bimonthly newsletter published by VermiCo, a company that sells worms but focuses mainly on educational seminars and workshops.

VermiCo
PO Box 2334
Grants Pass, OR 97528
www.vermico.com

## WEBSITES AND WORM FORUMS

**www.wormwoman.com**

The website of Mary Appelhof, leading worm educator and enthusiast. Books, worm bins, and—of course—worms are for sale.

**www.nrri.umn.edu/worms**

More information on the Minnesota Worm Watch program and the destruction of Minnesota's forests by worms.

**www.sarep.ucdavis.edu/worms**

Information on earthworms from the University of California at Davis, including species profiles for several common earthworms.

**www.soilfoodweb.com**

An interesting and informative site that offers soil tests and useful information about soil ecology; run by Dr. Elaine Ingham, who teaches workshops on organic soil management practices.

**www.livingsoil.com**

The website of Dr. Scott Subler's company, which sells bagged earthworm castings and tea.

**www.wormdigest.org**

Hosts a very active worm discussion forum with plenty of good advice on worms and worm composting. Check this site to find a worm grower in your area.

**www.happydranch.com**

Hosts an on-line discussion forum for Can-O-Worms users. Also sells worms, bins, books, and other items.

**www.squirmy-worms.com/discus/index.html**
Several worm forums on a variety of topics.

**www.unclejim.com/PAGES/bbs.shtml**
Wide-ranging discussions on everything worm related.

**www.recycleworks.co.uk/cgi-bin/ubbcgi/ultimatebb.cgi**
Includes an active worm forum for worm enthusiasts in the
United Kingdom.

## WORM BINS

New worm bins are coming on the market all the time. It's
a good idea to check some of the worm forums like
www.wormdigest.com for reviews of new models. Also, for any
size project, there is a homemade option. Small-scale worm bins
can be built in wooden boxes, plastic tubs, or garbage cans. Mid-
scale bins are often nothing more than large wooden boxes with
lids, or discarded freezers or refrigerators equipped with a few
modifications for safety and drainage. Even a large-scale worm
composting system can be made with windrows: long rows of
feedstock that worms work their way through, with a little shelter from the weather and a source of water to preserve ideal conditions for the worms.

You can buy a worm bin with or without worms included.
Many worm farmers sell a variety of bins in addition to worms,
worm books, and videos, and supplies such as bedding, ther-
mometers, and pH meters.

### Small Scale

There are several small, stacking worm bins on the market. These bins consist of three or four stacked trays with holes in the bottom of each tray; the worms move through the holes to reach the topmost layer of food. Eventually the bottom tray can be emptied and placed on top, allowing the process to begin again. Some popular brands include Can-O-Worms, Wriggly Wranch, and the Worm Factory. The bins are widely available from garden supply companies, worm growers, nurseries, and city worm-bin distribution programs.

Some worm farmers also sell wooden or plastic tubs that can be used as worm bins. They usually include drainage holes in the bottom and a few other special modifications. These bins are well suited for indoor use—for instance, in a classroom, under a kitchen sink, or in a cellar—but when it comes time to harvest the castings, it can be more difficult to separate the worms from the castings. Fortunately, this does not have to be done very often—worms are content to live in their castings for months at a time.

### Mid-Scale

**The Worm Wigwam** is one of the most popular mid-scale composters. It is three feet tall, three feet wide, and suitable for a restaurant, small grocery store, or small school. Available from EPM, Inc., PO Box 1295, Cottage Grove, OR 97424; visit www.wormwigwam.com or call 800-779-1709.

**The Eliminator 300** is manufactured by Happy D Ranch. It is built of wood and plexiglass and is appropriate for schools. Contact Happy D Ranch at PO Box 3001, Visalia, CA, 93278; visit www.happydranch.com, or call 559-738-9301.

### Large Scale

Large-scale bins are almost always built to custom specifications. They can be designed to work with existing equipment on a farm or in a factory. Often they are built on site inside a warehouse or barn. Some of the most popular models include the Worm Gin (352-485-1903), the Oregon Soil Corporation Reactor (503-557-9742), and the Vermitech (in Australia: +61 2 9261 4045 or www.vermitech.com).

# INDEX

6-26